Lecture Notes in Mathematics

A collection of informal reports and seminars
Edited by A. Dold, Heidelberg and B. Eckmann, Zürich

Series: Forschungsinstitut für Mathematik, ETH, Zürich · Adviser: K. Chandrasekharan

T0220585

97

Stephen U. Chase
Moss E. Sweedler
Cornell University, Ithaca, N.Y.

Hopf Algebras and Galois Theory

1969

Springer-Verlag Berlin · Heidelberg · New York

All rights reserved. No part of this book may be translated or reproduced in any form without written permission from Springer Verlag. © by Springer-Verlag Berlin · Heidelberg 1969
Library of Congress Catalog Card Number 75- 84143 · Printed in Germany. Title No. 3703

Preface and Acknowledgement

This volume consists of three chapters on separate but related topics, followed by several pages of footnotes which were inserted to correct and clarify certain points in the text.

Chapters I and III, written by Stephen U. Chase, are based upon a series of lectures given during the fall and winter of 1966-67 at the ETH Forschungsinstitut für Mathematik in Zürich. The author is indebted to the Forschungsinstitut and its director, Professor Beno Eckmann, for the gracious hospitality which he enjoyed there, and to the Alfred E. Sloan Foundation, ETH, and NSF GP-6484 for providing the necessary financial support.

Chapter II, which contains joint work of Stephen U. Chase and Moss E. Sweedler, was written during the fall of 1967 at Cornell University with the partial support of NSF GP-7945. Chapter III was subsequently revised so as to take advantage of some of the techniques of Chapter II. During this academic year the first-named author also received partial support from the Alfred E. Sloan Foundation.

The authors are indebted to their colleagues at Cornell and the Forschungsinstitut for many stimulating conversations on matters related to the material of these notes; in particular, to Jon Beck, Bill Lawvere, Fred Linton, Stephen Lichtenbaum, George Rinehart, and Miles Tierney.

Contents

I. Galois Objects

0. Introduction and Preliminaries

The notion of a Galois object in a category, as defined here, includes as special cases two familiar topics of study in algebra: Normal separable extensions of fields, and the extensions of various algebraic systems such as groups, associative algebras, and modules [11, Chapter XIV]. Thus it provides a convenient setting in which to discuss phenomena common to both situations. The purpose of these notes is to set forth certain facts regarding Galois objects, and then to analyze several special cases.

Section 1 consists primarily of characterizations and elementary properties of Galois G-objects, where G is a group in a given category. In Sections 2 and 3 we introduce for a category $\underline{A}$, a certain full subcategory $Ab(\underline{A})$ of the category of abelian groups in $\underline{A}$, and then construct an additive functor $\mathbf{X}$: $Ab(\underline{A}) \rightarrow Ab$, where Ab is the category of abelian groups. For J in $|Ab(\underline{A})|$, $\mathbf{X}(J)$ is the set of isomorphism classes of Galois J-objects satisfying an extra condition which is, in some sense, a generalization of the notion of a faithfully flat commutative algebra. This restriction is necessary for technical reasons; however, it turns out to be automatically satisfied in all of the special cases appearing in these notes.

Our construction of the functor $\mathbf{X}$ is, in essence, a generalization of Baer's origin construction of Ext(A,B) for abelian groups A and B [6]. That is, we work directly with the objects involved, rather than with representations of them by factor sets or cohomology classes. The treatment given here also bears a close resemblance to that of Harriso [22'],who constructed the functor $\mathbf{X}$ for the special case of abelian extensions of commutative rings.

In Section 4 we discuss and compute some examples of Galois objects in the categor of cocommutative coalgebras and commutative algebras over a commutative ring R. The Galo objects in the latter category are, modulo minor technical differences, the algebraic geometers' principal homogeneous spaces over affine group schemes. Specifically, if a ogroup A in the category of commutative R-algebras is a flat R-module, then the schemes hich are "formellement principal homogéne sous Spec (A) audessus de Spec (R)", in the ense of [41, Exposé 8], are precisely the affine schemes arising from Galois A-bjects which are flat R-modules.[1]

We refer the reader to Sections 6-12 for a more detailed treatment of Galois objects n the category of commutative R-algebras.

Section 5 contains some results on Galois objects in the category of algebras over triple ([16], [9], [7]); in particular, we observe that, in the relevant special cases, he Galois objects are precisely the familiar extensions of modules, groups, associative r commutative algebras, etc., and the groups X(J) are the groups of extensions arising rom these situations. We also apply the work of Beck [9] on triple cohomology to obtain cohomological description of the functor X which is, in our context, the analogue of rothendieck's generalization of Hilbert's Theorem 90 [20, Théorème 1]. We refer the eader to Sections 13-17 for a different cohomological description of the functor X in he context of commutative algebras.

It will be assumed here that the reader is familiar with what is common to the stan-dard expositions of the theory of categories ([17], [29], [27']), in particular with the ollowing concepts: Natural transformation and equivalence, pullback and pushout, product, imit, equalizer, adjoint functor, terminal object, dualization, and Yoneda's Lemma 27', p. 54]. The categories of sets and abelian groups will be denoted by Sets and Ab, espectively. The prefix "co-" adjoined to a word will denote the dual of the concept epresented by the word; this dual concept will usually not be explicitly defined.

Throughout the discussion the symbol $\underline{A}$ will, unless explicitly stated otherwise, lenote a category with finite products. The terminal object of $\underline{A}$ (which may be obtained y taking an empty product) will usually - but not always - be written 1. The class of bjects of $\underline{A}$ will be written $|\underline{A}|$. If X and Y are in $|\underline{A}|$, the set of morphisms from X to Y in $\underline{A}$ will be written $\underline{A}(X,Y)$, or simply (X,Y); an element of this set will often be re-ferred to as a map in $\underline{A}$, or an $\underline{A}$-map. If $\underline{B}$ is another category, the symbols $(\underline{A},\underline{B})$ will lenote the category of functors from $\underline{A}$ to $\underline{B}$. The dual category of $\underline{A}$ will be written $\underline{A}^{op}$.

If X is an object of $\underline{A}$, we shall denote by $(\underline{A},X)$ the category of which the objects

1 All footnotes see at the end on page 127

are all pairs (Y,p), with Y in $|\underline{A}|$ and $p\colon Y \to X$ an $\underline{A}$-map. A map $f\colon (Y_1,p_1) \to (Y_2,p_2)$ in $(\underline{A},X)$ is an $\underline{A}$-map $f\colon Y_1 \to Y_2$ with $p_2 f = p_1$. $X = (X,1_X)$ is a terminal object of $(\underline{A},X)$. We shall often write an object (Y,p) of (A,X) simply as Y, writing $p_Y = p$. The product in $(\underline{A},X)$ is the pullback over X, if the latter exists.

Given X as above, we obtain functors $\underline{A} \to \underline{A}$ and $\underline{A} \to (\underline{A},X)$ defined by $Y \to X \times Y$ and $Y \to (X \times Y,\text{projection on } X)$, respectively. We shall denote both of these by $X \times (\ \ )$; it will usually be clear from the context which is meant. In addition, we have the forgetful functor $(\underline{A},X) \to \underline{A}$ defined by $(Y,p) \to Y$, which is easily seen to be a left adjoint of $X \times (\ \ )\colon \underline{A} \to (\underline{A},X)$. In particular, the latter preserves limits.

The construction dual to $(\underline{A},X)$ will be written as $(X,\underline{A})$. Thus an object of $(X,\underline{A})$ is a pair (Y,i) with $i\colon X \to Y$ in $\underline{A}$, etc.

We next introduce a concept which will play an important role in these notes.

Definition 0.1. An object X of $\underline{A}$ will be called faithful if the following condition holds: Wherever $f\colon A \to B$ in $\underline{A}$ and $X \times f\colon X \times A \to X \times B$ is an isomorphism, then f is likewise an isomorphism.

Proposition 0.2. Let $f\colon X \to Y$ be an $\underline{A}$-map, with X faithful. Then Y is likewise faithful.

Proof. This is an immediate consequence of the equivalence $X \times (\ ) \approx X \times_Y (Y \times (\ ))$, the pullback $X \times_Y (\ )$ being constructed using the map f.

Corollary 0.3. Let X in $|\underline{A}|$ be such that $\underline{A}(1,X) \neq \emptyset$. Then X is faithful.

Proof. This follows from Proposition 0.2 and the obvious fact that 1 is faithful.

Corollary 0.4. Let X be a set, and $Y = (Y,p_Y)$ be in (Sets,X). Then Y is faithful if and only if $p_Y\colon Y \to X$ is onto, in which case there exists an (Sets,X)-map $s\colon X \to Y$.

Proof. $(\Leftarrow)$: If p_Y is onto, we use the Axiom of Choice to construct a map $s\colon (X,1_X) \to (Y,p_Y)$ in (Sets,X), after which we apply Corollary 0.3.
$(\Rightarrow)$: Let $X' = \text{Im}(p_Y) \subseteq X$ and $f\colon X' \to X$ be the inclusion map. We then obtain the map $f\colon (X',f) \to (X,1_X)$ in (Sets,X), giving rise to the map

$$(Y,p_Y) \times_X f\colon (Y,p_Y) \times_X (X',f) \to (Y,p_Y) \times_X (X,1_X) .$$

Applying the forgetful functor $(\text{Sets},X) \to \text{Sets}$, we obtain the following commutative diagram in Sets

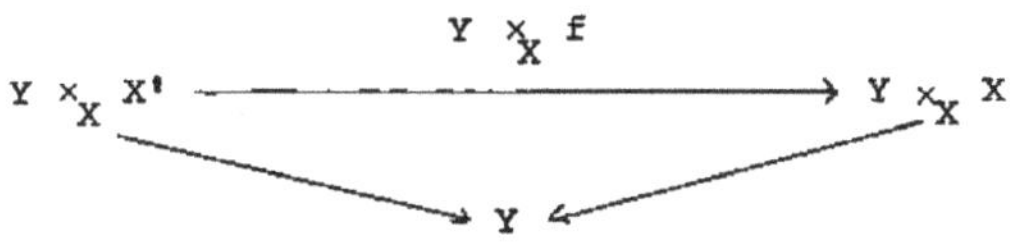

the unlabeled maps being the projections. It is easy to see that these are isomorphisms, and thus $Y \times_X f$ is an isomorphism. It is then clear that $(Y,p_Y) \times_X f$ is an isomorphism in (Sets,X). We may then conclude that f is an isomorphism, and p_Y is onto, completing the proof.

<u>Proposition 0.5.</u> Let X,Y be in $|\underline{A}|$, with Y faithful. Then $X \times Y$ is faithful in $(\underline{A},X)$.

Proof. Let $f: A \to B$ be an $(\underline{A},X)$-map such that $(X \times Y) \times_X f: (X\times Y)\times_X A \to (X\times Y)\times_X B$ is an isomorphism. Applying the forgetful functor $(\underline{A},X) \to \underline{A}$, we obtain that $Y \times f: Y \times A \to Y \times B$ is an isomorphism in $\underline{A}$, and hence the $\underline{A}$-map $f: A \to B$ is an isomorphism. But it is then clear that $f: A \to B$ is an isomorphism in $(\underline{A},X)$, and the proposition is proved.

Finally we shall have much use for the well-known concept of a group object in a category $\underline{A}$.

For brevity, we shall call a group object in $\underline{A}$ simply a group in $\underline{A}$; this is defined to be a 4-tuple $G = (G,\mu,\eta,\lambda)$ with G in $|\underline{A}|$ and $\mu: G \times G \to G$, $\eta: 1 \to G$, $\lambda: G \to G$ $\underline{A}$-maps such that the diagrams below commute.

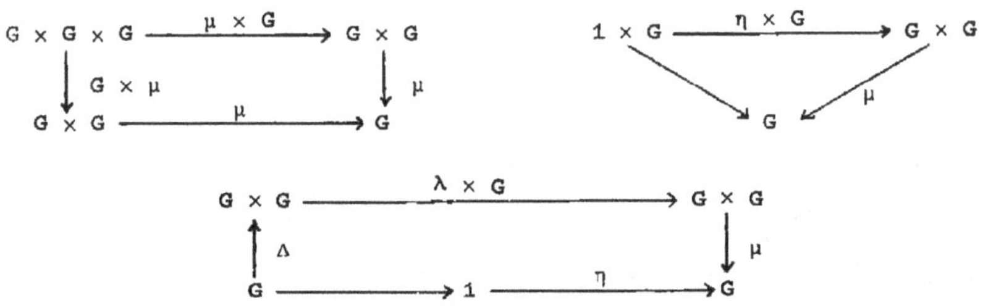

Here Δ is the diagonal map and the unlabeled maps are the obvious ones. μ,η,λ are called the multiplication, left identity, and left inverse of G, respectively. It is easy to see (e.g., by using properties of the Yoneda embedding $\underline{A} \to (\underline{A}^{op},\text{Sets})$) that η (resp. λ) is also, in the obvious sense of the term, a right identity (resp. right inverse) for G.

We shall usually denote the 4-tuple (G,μ,η,λ) by the symbol G alone. When explicit reference to μ is in order, we shall write μ_G, etc.

1. Elementary Properties of Galois Objects

Let $G = (G,\mu,\eta,\lambda)$ be a group in $\underline{A}$, as defined in Section 0. We define a G-object to be a pair (X,α), where X is in $|\underline{A}|$ and $\alpha: X \times G \to X$ in $\underline{A}$ such that the diagrams below commute.

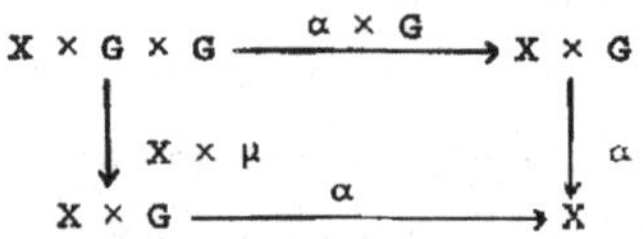

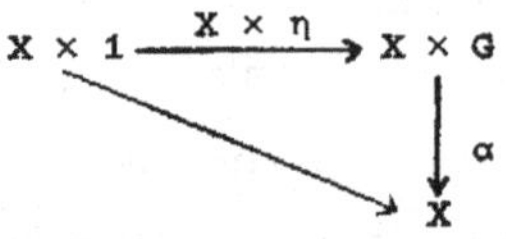

the unlabeled map being the projection. For brevity we shall often denote the pair (X,α) by X. If we need explicit reference to α we shall often write $\alpha = \alpha_X$. Note that G itself, with $\alpha_G = \mu$, is a G-object. An important role will be played by the category $\underline{A}^G$, the objects of which are all G-objects. A map $f: X \to Y$ in $\underline{A}^G$ is an $\underline{A}$-map such that $\alpha_Y(f \times G) = f\alpha_X$.

If X is a G-object, we define an $\underline{A}$-map $\gamma_X: X \times G \to X \times X$ by the product diagram:

(1.1)

$$\begin{array}{ccc}
& X \times G & \\
\swarrow & \downarrow \gamma_X & \searrow \alpha_X \\
X \longleftarrow & X \times X \longrightarrow & X
\end{array}$$

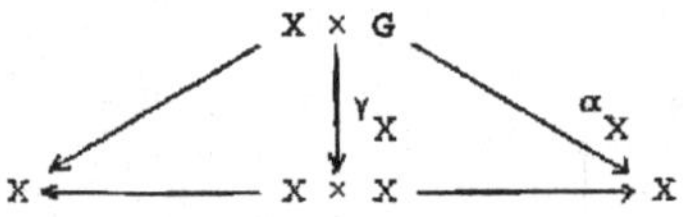

the unlabeled arrows denoting the projections.

<u>Definition 1.2.</u> Let G be a group in $\underline{A}$, and X be in $|\underline{A}^G|$. X will be called a <u>Galois G-object</u> if it is faithful (in the category $\underline{A}$) and the map $\gamma_X: X \times G \to X \times X$ is an isomorphism.

The following observation is appropriate at this point: Our assumption that $\underline{A}$ possess finite products could be relaxed to the requirement only that certain relevant products exist. That is, throughout our discussion the only products which we really need are those of the form $(\ ) \times G$ and $(\ ) \times X$, where G is a group in $\underline{A}$ and X is a G-object. Thus, for example, we could just as well have defined a group in $\underline{A}$ to be a 4-tuple (G,μ,η,λ) such that $(\ ) \times G$ exists and the conditions set forth at the end of Section 0 hold. This remark becomes pertinent when we consider the category $(\underline{A},X)$, with X a G-object. From time to time we shall wish to view $X \times G$ as a group in $(\underline{A},X)$ and $X \times X$ as an $X \times G$ -object. Although $(\underline{A},X)$ does not possess arbitrary finite products unless the pullback over X exists in $\underline{A}$, the relevant products $(\ ) \times_X (X \times G) \approx (\ ) \times G$ and $(\ ) \times_X (X \times X) \approx (\ ) \times X$ always exist.

The following useful remark, regarding the preservation of the above concepts by product-preserving functors, is easily checked.

Remark 1.3. Let $V: \underline{A} \to \underline{B}$ be a functor which preserves products. If G is a group in $\underline{A}$, then

(a) $V(G)$ is a group in $\underline{B}$, with operations obtained by applying V to their counterparts for G.

(b) If X is in $|\underline{A}^G|$, then $V(X)$ is in $|\underline{B}^{V(G)}|$, with $\alpha_{V(X)} = V(\alpha_X)$, $\gamma_{V(X)} = V(\gamma_X)$. If $f: X \to Y$ is an $\underline{A}^G$-map, then $V(f): V(X) \to V(Y)$ is a $\underline{B}^{V(G)}$-map.

(c) If X is a Galois G-object and $V(X)$ is faithful, then $V(X)$ is a Galois $V(G)$-object.

Corollary 1.4. Let G be a group in $\underline{A}$, and Y be in $|\underline{A}^G|$. Then

(a) $\gamma_Y: Y \times G \to Y \times Y$ is a map in the category $(\underline{A},Y)^{Y \times G}$.

(b) If X is in $|\underline{A}|$, and Y is a Galois G-object, then $X \times Y$ is a Galois $X \times G$ -object in the category $(\underline{A},X)$.

Proof. Note first that both statements make sense, in view of Remark 1.3 and the fact, noted in Section 0, that the functors $X \times (\ ): \underline{A} \to (\underline{A},X)$, etc., preserve products. Hence $X \times G$ has the structure of a group in $(\underline{A},X)$, $X \times Y$ has the structure of an $X \times G$ -object, and so forth. (a) then follows from an easy computation, whereas (b) is a consequence of Remark 1.3. (c) and Proposition 0.5.

Next we scrutinize the special case $\underline{A}$ = Sets. Of course, a group G in Sets is an ordinary group, and a G-object X is simply a set on which G operates from the right. We write this operation as justaposition; i.e., we write $x\sigma = \alpha_X(x,\sigma)$ for x in X, σ in G. $\gamma_X: X \times G \to X \times X$ then satisfies the formula

(1.5) $$\gamma_X(x,\sigma) = (x,x\sigma) \qquad (x \text{ in } X,\ \sigma \text{ in } G)$$

Lemma 1.6. Let G be a group, and $X \neq \emptyset$ be in $Sets^G$. Then the following conditions are equivalent:

(a) $\gamma_X: X \times G \to X \times X$ is an isomorphism.

(b) $X \times X \approx X \times G$ in $(Sets,X)^{X \times G}$.

(c) $X \approx G$ in $Sets^G$.

Proof. That (a) implies (b) is an easy consequence of Corollary 1.4(a). The remaining implications follow from routine computations which we omit.

Corollary 1.7. If G is a group and X is in $|Sets^G|$, then X is a Galois G-object if and only if $X \approx G$ in $Sets^G$.

Proof. This is an immediate consequence of Lemma 1.6 and the trivial fact that X is faithful if and only if $X \neq \emptyset$.

Theorem 1.8. Let G be a group in $\underline{A}$. Then the following statements are equivalent for any Y in $|\underline{A}^G|$:

(a) $\gamma_Y: Y \times G \to Y \times Y$ is an isomorphism.

(b) $Y \times Y \approx Y \times G$ in $(\underline{A},Y)^{Y \times G}$.

(c) For any X in $|\underline{A}|$, either $(X,Y) = \emptyset$ or $(X,Y) \approx (X,G)$ in $Sets^{(X,G)}$.

In particular, if Y is a Galois G-object, then it satisfies all of the above conditions.

Proof. Note first that (c) makes sense, in view of Remark 1.3 and the fact that the functor $(X,-): \underline{A} \to Sets$ preserves products.

(a) $\to$ (b): Use Corollary 1.4(a).

(b) $\to$ (c): Let X in $\underline{A}$ be such that $(X,Y) \neq \emptyset$. Applying Remark 1.3 with $V = (X,-): \underline{A} \to Sets$, we obtain easily that

$$(X,Y) \times (X,Y) \approx (X,Y) \times (X,G)$$

in the category $(Sets, (X,Y))^{(X,Y) \times (X,G)}$. Thus, by Lemma 1.6, $(X,Y) \approx (X,G)$ in $Sets^{(X,G)}$, establishing (c).

(c) $\to$ (a): Let X be in $|\underline{A}|$. If $(X,Y) = \emptyset$, then $(X,Y \times G) = \emptyset = (X,Y \times Y)$, and thus

$$(X,\gamma_Y): (X,Y \times G) \to (X,Y \times Y)$$

is an isomorphism. Suppose on the other hand that $(X,Y) \neq \emptyset$. Note that $(X,G) \neq \emptyset$, since it contains the map

$$X \to 1 \xrightarrow{\eta} G$$

The commutative diagram below then follows easily from Remark 1.3

$$\begin{array}{ccc} (X, Y \times G) & \xrightarrow{(X,\gamma_Y)} & (X, Y \times Y) \\ \| & & \| \\ (X,Y) \times (X,G) & \xrightarrow{\gamma_{(X,Y)}} & (X,Y) \times (X,Y) \end{array}$$

the vertical isomorphisms being the obvious ones. Since $\gamma_{(X,Y)}$ is an isomorphism by Lemma 1.6, we obtain that (X,γ_Y) is an isomorphism in this case, too. (a) then follows from the Yoneda Lemma, completing the proof.

<u>Corollary 1.9.</u> If G is a group in $\underline{A}$, then G is a Galois G-object.

Proof. The existence of $\eta\colon 1 \to G$ and Corollary 0.3 guarantee that G is faithful. But clearly $G \times G \approx G \times G$ in $(\underline{A},G)^{G \times G}$, and the corollary then follows from Theorem 1.8.

<u>Definition and Remark 1.10.</u> If G is as above, then a Galois G-object X will be called <u>trivial</u> if $X \approx G$ in $\underline{A}^G$. It follows immediately from Lemma 1.6 that, in the special case $\underline{A}$ = Sets, every Galois G-object is trivial.

We turn next to maps of Galois objects.

<u>Lemma 1.11.</u> Let G be a group in $\underline{A}$, Y be a Galois G-object, and $f\colon G \to Y$ be an $\underline{A}^G$-map. Then f is an isomorphism.

Proof. Let X be in $|\underline{A}|$; then $(X,G) \neq \emptyset$, and thus $(X,Y) \neq \emptyset$. Then $(X,Y) \approx (X,G)$ in $\mathrm{Sets}^{(X,G)}$, by Theorem 1.8. An easy computation then establishes that the map $(X,f)\colon (X,G) \to (X,Y)$ is an isomorphism. We apply the Yoneda Lemma to complete the proof.

<u>Theorem 1.12.</u> Let G be a group in $\underline{A}$, X and Y be Galois G-objects, and $f\colon X \to Y$ be an $\underline{A}^G$-map. Then f is an isomorphism.

Proof. We have from Remark 1.3 and Corollary 1.4 that $X \times Y$ is a Galois $X \times G$ -object in the category $(\underline{A},X)$, and $X \times f\colon X \times X \to X \times Y$ is an $(\underline{A},X)^{X \times G}$-map. Also, $X \times X$ is trivial by Theorem 1.8, and so $X \times f$ is an isomorphism in $(\underline{A},X)$, by Lemma 1.11. But then the $\underline{A}$-map $X \times f\colon X \times X \to X \times Y$ is likewise an isomorphism, whence the theorem follows from the fact that X is faithful.

<u>Corollary 1.13.</u> View 1 as a group in $\underline{A}$, with the obvious structure. Then every Galois 1-object is trivial.

Proof. If X is a Galois 1-object, the map $X \to 1$ is easily seen to be in $\underline{A}^1$. Apply Theorem 1.12.

Corollary 1.14. Let X be a Galois G-object, G a group in $\underline{A}$. Then X is trivial if and only if there exists an $\underline{A}$-map $s\colon 1 \to X$.

Proof. If X is trivial, then the identity $\eta\colon 1 \to G$ of G gives rise to an $\underline{A}$-map $s\colon 1 \to X$. Conversely, given $s\colon 1 \to X$, define $f\colon G \to X$ by the commutative diagram

$$\begin{array}{ccc} 1 \times G & \xrightarrow{\;s \times G\;} & X \times G \\ \| & & \big\downarrow{\scriptstyle \alpha_X} \\ G & \xrightarrow{\quad f \quad} & X \end{array}$$

the unlabeled isomorphism being the projection. One checks easily that f is an $\underline{A}^G$-map, hence an isomorphism by Theorem 1.12. That is, X is trivial, completing the proof.

Before passing to other matters, we give another useful criterion for a G-object to be Galois.

Theorem 1.15. If G is a group in $\underline{A}$, then the following statements are equivalent for any Y in $|\underline{A}^G|$:

(a) Y is a Galois G-object.

(b) $X \times Y \approx X \times G$ in $(\underline{A},X)^{X \times G}$ for some faithful X in $|\underline{A}|$.

(c) Y is faithful, and there exists a faithful X in $|\underline{A}|$ such that $X \times Y$ is a Galois $X \times G$ -object in $(\underline{A},X)$.

Proof. (a) $\Rightarrow$ (b): Set X = Y and use Theorem 1.8.

(b) $\Rightarrow$ (c): Select X as in (b). Then $X \times Y$ is a Galois $X \times G$ -object, by Corollary 1.9. Furthermore, the $\underline{A}$-map

$$X \xrightarrow{(1_X,\eta)} X \times G \xrightarrow{\;\approx\;} X \times Y$$

guarantees, via Proposition 0.2, that $X \times Y$ is faithful in $\underline{A}$, whence the projection map $X \times Y \to Y$ then ensures, for the same reason, that Y is faithful. This establishes (c).

(c) $\Rightarrow$ (a): Let X in $|\underline{A}|$ be as in (c). Remark 1.3 and Corollary 1.4 then give easily the following commutative diagram in $(\underline{A},X)$

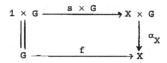

$$\begin{array}{ccc} X \times (Y \times G) & \xrightarrow{X \times \gamma_Y} & X \times (Y \times Y) \\ \| & & \| \\ (X \times Y) \times_X (X \times G) & \xrightarrow{\gamma_{X \times Y}} & (X \times Y) \times_X (X \times Y) \end{array}$$

the vertical isomorphisms being the obvious ones. (c) then guarantees that $X \times \gamma_Y$ is an isomorphism in $(\underline{A},X)$. But then $X \times \gamma_Y$: $X \times (Y \times G) \to X \times (Y \times Y)$ is an isomorphism in A, and therefore γ_Y is likewise because X is faithful. Since Y is faithful, it is then a Galois G-object. This establishes (a) and completes the proof of the theorem.

The statements in the example below are easy consequences of the preceding discussion.

Example 1.16. Let G be a group (in Sets), and A be in $|\text{Sets}^G|$. In this example we shall write $\alpha_A(a,\sigma) = a^\sigma$ for a in A, σ in G; then of course

$$a^{\sigma\tau} = (a^\sigma)^\tau$$

for σ,τ in G. Suppose now that A is a group in Sets^G; then A is an ordinary group such that the equation below holds for all a,b in A and σ in G

$$(ab)^\sigma = a^\sigma b^\sigma$$

We set $\underline{A} = \text{Sets}^G$, and consider an object P of $\underline{A}^A$. Then P is a set on which A operates from the right via the equation $ua = \alpha_P(u,a)$ for u in P, a in A. Moreover, the equation below holds for all u in P, a in A, σ in G

$$(ua)^\sigma = u^\sigma a^\sigma$$

Finally, it is easily verified that P is a Galois A-object if and only if $P \neq \emptyset$ and A operates simply and transitively on P; i.e., P is a "principal homogeneous space over A" in the sense of Serre [38,p.I-58].

2. Adjointness and the Functor E

In this section and the next, we shall assume that the category $\underline{A}$ possesses finit products and coequalizers [29 , p.8.]. Our aim is to make the collection of isomorphism classes of Galois G-objects (G a group in $\underline{A}$) into a pointed set which is functorial in G. To this end, we first introduce into our setting certain necessary restrictions.

We shall call a diagram

$$(2.1) \qquad A \underset{g}{\overset{f}{\rightrightarrows}} B \xrightarrow{h} C \qquad \text{in } \underline{A}$$

a coequalizer diagram (c.d.) if h is the coequalizer of the pair of maps (f,g).

Lemma 2.2. If X is in $\underline{A}$, then $(\underline{A},X)$ possesses coequalizers. Furthermore, a diagram

$$(2.3) \qquad A \underset{g}{\overset{f}{\rightrightarrows}} B \xrightarrow{h} C \qquad \text{in } (\underline{A},X)$$

is a c.d. if and only if its image in $\underline{A}$ under the forgetful functor $(\underline{A},X) \to \underline{A}$ is like wise.

Proof. In this proof (as later) we shall, whenever convenient, denote an object $(\underline{A},X)$, and its image in $\underline{A}$ under the forgetful functor, by the same symbol; the meaning will usually be clear from the context.

Given the c.d. (2.3), we may prove that its image in $\underline{A}$ is likewise a c.d. either an easy direct argument, or by using the fact that a left adjoint preserves colimits.

Conversely, assume given (2.3) such that its image (2.1) in $\underline{A}$ is a c.d. If $\beta : B \to D$ in $(\underline{A},X)$ is such that $\beta f = \beta g$, then there exists a unique $\underline{A}$-map $\alpha : C \to D$ such that $\beta = \alpha h$. We then have the following commutative diagram in $\underline{A}$

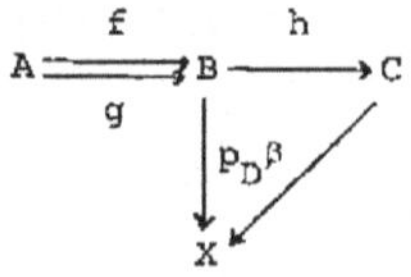

where the unlabeled map is either p_C or $p_D\alpha$. The definition of c.d. then guarantees that $p_C = p_D\alpha$, in which case $\alpha: C \to D$ is an $(\underline{A},X)$-map and $\beta = \alpha h$ in $(\underline{A},X)$. In order to complete the proof that (2.3) is a c.d., we need only show that α is the unique $(\underline{A},X)$-map $C \to D$ whose composition with h is β. But if also $\alpha': C \to D$ is an $(\underline{A},X)$-map and $\alpha' h = \beta$, then also $\alpha' h = \beta$ in $\underline{A}$, in which case $\alpha' = \alpha$ because (2.1) is a c.d.

Finally, given $f,g: A \to B$ in $(\underline{A},X)$, we obtain a c.d. (2.1) in $\underline{A}$. Since $p_B f = p_A = p_B g$, there exists a unique $\underline{A}$-map $p_C: C \to X$ with $p_C h = p_B$, in which case (2.3) is a c.d., and the proof is complete.

Definition 2.4. Let X be in $|\underline{A}|$.

(a) X will be called coflat if it preserves coequalizer diagrams in the following sense: Given a c.d. (2.1) in $\underline{A}$, then the diagram

(2.5) $$X \times A \underset{X \times g}{\overset{X \times f}{\rightrightarrows}} X \times B \xrightarrow{X \times h} X \times C \quad \text{in } \underline{A}$$

is likewise a c.d.

(b) X will be called faithfully coflat (f.c.) if it is coflat and faithful.

We shall show later that, in the appropriate setting, the dual of Definition 2.4 reduces to the notions of flatness and faithful flatness of [10, Chapter I.].

We omit the easy verification of the properties, set forth below, of coflat and faithfully coflat objects.

Remark 2.6. The following statements are equivalent for X in $|\underline{A}|$

(a) X is f.c.

(b) X preserves and reflects coequalizers; i.e., a diagram (2.1) in $\underline{A}$, with $hf = hg$, is a c.d. if and only if (2.5) is likewise.

Proposition 2.7. Let X,Y be in $|\underline{A}|$.

(a) If X and Y are coflat, then so is $X \times Y$.

(b) If X an f.c., then $X \times Y$ is coflat (f.c.) if and only if Y is coflat (f.c.).

(c) If Y is coflat (f.c.) in $\underline{A}$, then $X \times Y$ is coflat (f.c.) in $(\underline{A},X)$.

Remark 2.8. If G is a coflat group in $\underline{A}$ (i,e., a group in $\underline{A}$ which is coflat as an object of $\underline{A}$) then G is f.c.

We return now to our discussion of Galois objects. Let $\varphi\colon G \to H$ be a homomorphism of groups in $\underline{A}$; i.e., φ is an $\underline{A}$-map such that the diagram below commutes

$$\begin{array}{ccc} G \times G & \xrightarrow{\varphi \times \varphi} & H \times H \\ \downarrow{\scriptstyle \mu_G} & & \downarrow{\scriptstyle \mu_H} \\ G & \xrightarrow{\varphi} & H \end{array}$$

with μ_G the multiplication of G, etc. Given a Galois G-object Y, we wish to construct from it and φ a Galois H-object, in a natural way. To this end we define a functor $\underline{A}^{\varphi}\colon \underline{A}^H \to \underline{A}^G$ as follows. If X is in $|\underline{A}^H|$, then $\underline{A}^{\varphi}(X) = (X, \alpha_{\underline{A}^{\varphi}(X)})$, with $\alpha_{\underline{A}^{\varphi}(X)}$ the composition

$$X \times G \xrightarrow{X \times \varphi} X \times H \xrightarrow{\alpha_X} X$$

If $f\colon X \to X'$ in $\underline{A}^H$, then $\underline{A}^{\varphi}(f) = f$.

<u>Theorem 2.9.</u> Let $\varphi\colon G \to H$ be a homomorphism of groups in $\underline{A}$, with H coflat. Then th functor $\underline{A}^{\varphi}\colon \underline{A}^H \to \underline{A}^G$ possesses a left adjoint $\tilde{\varphi}\colon \underline{A}^G \to \underline{A}^H$. If X is in $|\underline{A}^G|$, then the diagram below is a c.d. in $\underline{A}$

(2.10)
$$X \times G \times H \underset{\alpha_X \times H}{\overset{\omega_{X,\varphi}}{\rightrightarrows}} X \times H \xrightarrow{\varrho_{X,\varphi}} \tilde{\varphi}(X)$$

where $\omega_{X,\varphi}$ and $\varrho_{X,\varphi}$ are the compositions

$$X \times G \times H \xrightarrow{X \times \varphi \times H} X \times H \times H \xrightarrow{X \times \mu_H} X \times H$$

$$X \times H \xrightarrow{\theta_X \times H} \tilde{\varphi}(X) \times H \xrightarrow{\alpha_{\tilde{\varphi}(X)}} \tilde{\varphi}(X)$$

respectively, and $\theta_X\colon X \to \underline{A}^{\varphi}\tilde{\varphi}(X)$ is the $\underline{A}^G$-map given by the adjointness transformation $\theta\colon 1_{\underline{A}^G} \to \underline{A}^{\varphi}\tilde{\varphi}$. In particular, if $f\colon X \to \underline{A}^{\varphi}(Y)$ is an $\underline{A}^G$-map (Y in $|\underline{A}^H|$), then the corresponding $\underline{A}^H$-map $\bar{f}\colon \tilde{\varphi}(X) \to Y$, arising from adjointness, renders the diagram below commutative

$$\begin{array}{ccc} X \times H & \xrightarrow{\varrho_{X,\varphi}} & \tilde{\varphi}(X) \\ \Big\downarrow f \times H & & \Big\downarrow \overline{f} \\ Y \times H & \xrightarrow{\alpha_Y} & Y \end{array}$$

Proof. This is a consequence, via the remarks at the beginning of Section 1, of a theorem on triples which has been proved in unpublished work of Beck, Lawvere, and Linton. A direct proof can be given of the special case considered here. One defines $\tilde{\varphi}(X)$ (as an object of $\underline{A}$) by the c.d. (2.10), which exists by assumption. One then applies the functor $(\ ) \times H$ to (2.10) and uses the coflatness of H to obtain the structure of an H-object on $\tilde{\varphi}(X)$ such that $\varrho_{X,\varphi}$ is an $\underline{A}^H$-map. The proof of adjointness is then routine.

Remark 2.11. Let $\underline{B}$ be a category with finite products and coequalizers, and $V: \underline{A} \to \underline{B}$ be a functor which preserves products. Let $\varphi: G \to H$ be a homomorphism of groups in $\underline{A}$, with H coflat. If X is an object of $\underline{A}^G$, then the adjointness $\underline{A}^G$-map $\theta_X: X \to \underline{A}^{\varphi}\tilde{\varphi}(X)$ gives rise, via Remark 1.3, to the $\underline{B}^{V(G)}$-map

$$V(\theta_X): V(X) \to V\{\underline{A}^{\varphi}\tilde{\varphi}(X)\} = \underline{B}^{V(\varphi)}\{V(\tilde{\varphi}(X))\}$$

If V(H) is a coflat group in $\underline{B}$, then adjointness gives rise, via Theorem 2.9, to a $\underline{B}^{V(H)}$-map

$$\overline{V(\theta_X)}: \widetilde{V(\varphi)}(V(X)) \to V(\tilde{\varphi}(X)) \tag{2.12}$$

Lemma 2.13. Let V and φ be as in Remark 2.11. If V preserves products and coequalizers, then the map (2.12) is an isomorphism.

Proof. We apply V to (2.10) to obtain the c.d.

$$V(X) \times V(G) \times V(H) \underset{\alpha_{V(X)} \times V(H)}{\overset{\omega_{V(X),V(\varphi)}}{\rightrightarrows}} V(X) \times V(H) \xrightarrow{V(\varrho_{X,\varphi})} V(\tilde{\varphi}(X)) \tag{2.14}$$

But we also have the c.d.

$$V(X) \times V(G) \times V(H) \underset{\alpha_{V(X)} \times V(H)}{\overset{\omega_{V(X),V(\varphi)}}{\rightrightarrows}} V(X) \times V(H) \xrightarrow{\varrho_{V(X),V(\varphi)}} \widetilde{V(\varphi)}(V(X)) \tag{2.15}$$

Now scrutinize the diagram below in the category $\underline{B}$

(2.16)
$$\begin{array}{ccc} V(X) \times V(H) & \xrightarrow{\ \varrho_{V(X),V(\varphi)}\ } & \widetilde{V(\varphi)}(V(X)) \\ \downarrow{\scriptstyle V(\theta_X) \times V(H)} & \searrow{\scriptstyle V(\varrho_{X,\varphi})} & \downarrow{\scriptstyle \overline{V(\theta_X)}} \\ V(\tilde{\varphi}(X)) \times V(H) & \xrightarrow[\ \alpha_{V(\tilde{\varphi}(X))}\ ]{} & V(\tilde{\varphi}(X)) \end{array}$$

The lower triangle commutes, in view of the definition of $\varrho_{X,\varphi}$ and the fact that V preserves products. Furthermore, the square commutes in view of the last statement of Theorem 2.9. Thus the upper triangle commutes as well. The c.d.'s (2.14) and (2.15) then guarantee that $\overline{V(\theta_X)}$ is an isomorphism, completing the proof.

Corollary 2.17. (a) If X is in $\underline{A}$, then $(\underline{A},X)$ possesses coequalizers.

(b) If $\varphi: G \to H$ is a homomorphism of groups in $\underline{A}$, with H coflat, then $X \times H$ is a coflat group in $(\underline{A},X)$ and the $(\underline{A},X)^{X \times H}$-map

$$\overline{X \times \theta_Y}: (\widetilde{X \times \varphi})(X \times Y) \to X \times \tilde{\varphi}(Y)$$

is an isomorphism for any Y in $|\underline{A}^G|$.

Proof. These assertions are immediate consequences of Lemma 2.13, Proposition 2.7, and Lemma 2.2.

Lemma 2.18. Let $\varphi: G \to H$ be a homomorphism of groups in $\underline{A}$, with H coflat. Then $\tilde{\varphi}(G) \approx H$ in $\underline{A}^H$.

Proof. Consider the diagram

(2.19)
$$G \times G \times H \underset{\mu_G \times H}{\overset{\omega_{G,\varphi}}{\rightrightarrows}} G \times H \xrightarrow{\ \xi\ } H$$

where ξ is the composition

$$G \times H \xrightarrow{\ \varphi \times H\ } H \times H \xrightarrow{\ \mu_H\ } H$$

If $f: G \times H \to Y$ is an $\underline{A}$-map such that $f\omega_{G,\varphi} = f(\mu_G \times H)$, then define $g: H \to Y$ to be the composition

$$H = 1 \times H \xrightarrow{\eta_G \times H} G \times H \xrightarrow{f} Y$$

One checks easily that g is the unique $\underline{A}$-map from H to Y such that $g\xi = f$; thus (2.19) is a c.d. in $\underline{A}$. We may then apply Theorem 2.9 to obtain an $\underline{A}$-isomorphism $H \to \tilde{\varphi}(G)$, which is in fact the composition

$$H = 1 \times H \xrightarrow{\eta_G \times H} G \times H \xrightarrow{\varrho_{G,\varphi}} \tilde{\varphi}(G)$$

and is hence an $\underline{A}^H$-map, since $\varrho_{G,\varphi}$ is. The result follows.

<u>Theorem 2.20.</u> Let φ: $G \to H$ be a homomorphism of coflat groups in $\underline{A}$, and let X be a coflat Galois G-object. Then $\tilde{\varphi}(X)$ is a coflat Galois H-object.

Proof. As in the preceding discussion, let θ: $1_{\underline{A}^G} \to \underline{A}^{\varphi}\tilde{\varphi}$ be the adjointness transformation. Then the $\underline{A}$-map θ_X: $X \to \underline{A}^{\varphi}\tilde{\varphi}(X) = \tilde{\varphi}(X)$ shows, via Proposition 0.2, that $\tilde{\varphi}(X)$ is faithful.

We have, by Theorem 1.8, an $(\underline{A},X)^{X \times G}$-isomorphism $X \times X \approx X \times G$. Also, we obtain from Corollary 2.17 and Theorem 2.9 that the functor

$$\widetilde{X \times \varphi}\colon (\underline{A},X)^{X \times G} \to (\underline{A},X)^{X \times H}$$

arising from the homomorphism $X \times \varphi$: $X \times G \to X \times H$ of groups in $(\underline{A},X)$, is well-defined. Applying this functor to the preceding isomorphism, we obtain an isomorphism $(\widetilde{X \times \varphi})(X \times X) \approx (\widetilde{X \times \varphi})(X \times G)$ in $(\underline{A},X)^{X \times H}$. Since X and H are coflat, we may then use Corollary 2.17 and Lemma 2.18 to obtain the following chain of $(\underline{A},X)^{X \times H}$-isomorphisms

$$(2.21) \qquad X \times \tilde{\varphi}(X) \approx (\widetilde{X \times \varphi})(X \times X) \approx (\widetilde{X \times \varphi})(X \times G) \approx X \times \tilde{\varphi}(G) \approx X \times H$$

Since X is faithful, we may then apply Theorem 1.15 to conclude that $\tilde{\varphi}(X)$ is a Galois H-object. That $\tilde{\varphi}(X)$ is f.c. follows easily, via Proposition 2.7, from the fact X and H are both f.c. This completes the proof of the theorem.

<u>Definition and Remarks 2.22.</u> If G is a group in $\underline{A}$, we shall denote by E(G) the collection of $\underline{A}^G$-isomorphism classes of <u>coflat</u> Galois G-objects. Let Gp($\underline{A}$) be the ca-

tegory of which the objects are all groups G in $\underline{A}$ such that

(a) G is coflat.

(b) E(G) is a set.

and the maps are all homomorphisms of such groups. If G is in $|Gp(\underline{A})|$, then E(G) is a pointed set with base-point cl(G), where "cl( )" means "$\underline{A}^G$-isomorphism class of ( )". If $\varphi: G \to H$ in $Gp(\underline{A})$, then Lemma 2.18 and Theorem 2.20 guarantee that the formula $E(\varphi)(cl(X)) = cl(\tilde{\varphi}(X))$ defines a map $E(\varphi): E(G) \to E(H)$ of pointed sets. Finally, if $\varphi: G \to H$ and $\psi: H \to J$ in $Gp(\underline{A})$, then one checks easily that $\underline{A}^{\psi\varphi} = \underline{A}^{\varphi}\underline{A}^{\psi}: \underline{A}^J \to \underline{A}^G$. Thus, by the uniqueness of adjoints, we obtain a natural equivalence of functors $\widetilde{\psi\varphi} \approx \tilde{\psi}\tilde{\varphi}: \underline{A}^G \to \underline{A}^J$, which gives rise to the equality $E(\psi\varphi) = E(\psi)E(\varphi): E(G) \to E(J)$. We conclude that E is a functor from $Gp(\underline{A})$ to the category of pointed sets.

We end this section with a property of the map $\theta_X: X \to \underline{A}^{\varphi}\tilde{\varphi}(X) = \tilde{\varphi}(X)$ which will be needed in Section 16.

<u>Proposition 2.23.</u> Let $\varphi: G \to H$ be a homomorphism of coflat groups in $\underline{A}$.

(a) If X is a G-object, then $\alpha_{\tilde{\varphi}(X)}(\theta_X \times \varphi) = \theta_X\alpha_X: X \times G \to \tilde{\varphi}(X)$.

(b) If X is a coflat Galois G-object, Y is a coflat Galois H-object, and $f: X \to Y$ in $\underline{A}$ such that $\alpha_Y(f \times \varphi) = f\alpha_X: X \times G \to Y$, then there exists a unique $\underline{A}^H$-<u>isomorphism</u> $\bar{f}: \tilde{\varphi}(X) \stackrel{\approx}{\to} Y$ such that $\bar{f}\theta_X = f: X \to Y$.

Proof. The equality in (a) simply expresses the fact that $\theta_X: X \to \underline{A}^{\varphi}\tilde{\varphi}(X)$ is a map in the category $\underline{A}^G$. The first equality in (b) expresses the fact that $f: X \to \underline{A}^{\varphi}(Y)$ is an $\underline{A}^G$-map, whence the desired $\underline{A}^H$-map $\bar{f}: \tilde{\varphi}(X) \to Y$ is given by adjointness and is an isomorphism by Theorems 2.20 and 1.12.

3. Products of Galois Objects and the Functor $\mathcal{X}$

In this section we shall assume, unless stated otherwise, that the categories $\underline{A}$ with which we deal possess finite products and coequalizers. We shall denote by $Ab(\underline{A})$ the full subcategory of $Gp(\underline{A})$, of which the objects are all objects of $Gp(\underline{A})$ which are abelian. In this section we make $E(J)$, for J in $|Ab(\underline{A})|$, into an abelian group, thereby obtaining a functor $\mathcal{X}$: $Ab(\underline{A}) \to Ab$.

Lemma 3.1. Let G,H be groups in $\underline{A}$, X be a Galois G-object, and Y be a Galois H-object. Then $W \times X \approx W \times G$ in $(\underline{A},W)^{W \times G}$ and $W \times Y \approx W \times H$ in $(\underline{A},W)^{W \times H}$, with $W = X \times Y$.

Proof. The result follows from Theorem 1.8, the commutativity (up to isomorphism) of the diagram below

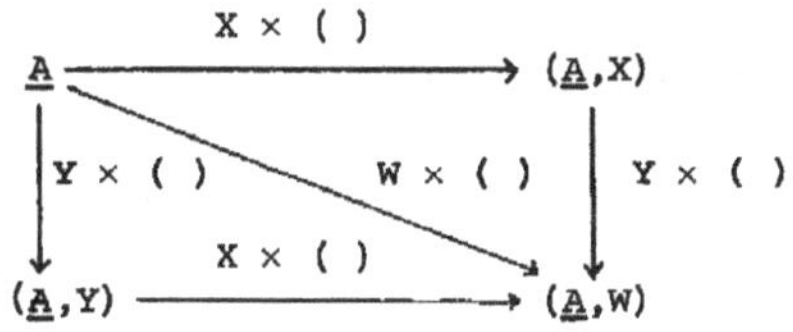

and the fact that the functors involved preserve the relevant data (Remark 1.3.).

Proposition 3.2. Let G,H,X,Y be as in Lemma 3.1. Then $X \times Y$ is a Galois $G \times H$ - object.

Proof. That $W = X \times Y$ is faithful follows trivially from the fact that X and Y are. Since $W \times (\ ): \underline{A} \to (\underline{A},W)$ preserves products, we obtain from Lemma 3.1 the chain of isomorphisms

$$W \times W \approx (W \times X) \times_W (W \times Y) \approx (W \times G) \times_W (W \times H) \approx W \times (G \times H) \quad \text{in } (\underline{A},W)^{W \times (G \times H)}$$

The result then follows from Theorem 1.8.

Definition and Remark 3.3. Let $\varphi_i: G_i \to H_i$ be homomorphisms of coflat groups in $\underline{A}$, and X_i be in $|\underline{A}^{G_i}|$ $(i = 1,2)$. We then have the commutative diagrams

(3.4)
$$\begin{array}{ccc} G_1 \times G_2 & \xrightarrow{\varphi_1 \times \varphi_2} & H_1 \times H_2 \\ \Big\downarrow \nu_i & & \Big\downarrow \pi_i \\ G_i & \xrightarrow{\varphi_i} & H_i \end{array}$$

with ν_i, π_i the projections. Now consider the $\underline{A}^{G_1 \times G_2}$-maps defined to be the compositions

(3.5)
$$X_1 \times X_2 \xrightarrow{p_i} \underline{A}^{\nu_i}(X_i) \xrightarrow{\underline{A}^{\nu_i}(\theta^i_{X_i})} \underline{A}^{\nu_i}\underline{A}^{\varphi_i}(\tilde{\varphi}_i(X_i)) = \underline{A}^{\varphi_1 \times \varphi_2}\,\underline{A}^{\pi_i}\,(\tilde{\varphi}_i(X_i))$$

with p_i the projection (easily seen to be an $\underline{A}^{G_1 \times G_2}$-map), $\theta^i\colon 1_{\underline{A}^{G_i}} \to \underline{A}^{\varphi_i}\tilde{\varphi}_i$ the adjointness transformation, and the last equality an easy consequence of the commutativi of (3.4). (3.5) and adjointness then give $\underline{A}^{H_1 \times H_2}$-maps

$K_{X_i}\colon (\widetilde{\varphi_1 \times \varphi_2})(X_1 \times X_2) \to \underline{A}^{\pi_i}\,(\tilde{\varphi}_i(X_i))$. Finally, we obtain an $\underline{A}$-map

$L_{X_1,X_2}\colon (\widetilde{\varphi_1 \times \varphi_2})(X_1 \times X_2) \to \tilde{\varphi}_1(X_1) \times \tilde{\varphi}_2(X_2)$ from the following product diagram in $\underline{A}$

$$\begin{array}{ccccc} & & (\widetilde{\varphi_1 \times \varphi_2})(X_1 \times X_2) & & \\ & \swarrow K_{X_1} & \Big\downarrow L_{X_1,X_2} & K_{X_2} \searrow & \\ \underline{A}^{\pi_1}(\tilde{\varphi}_1(X_1)) = \tilde{\varphi}_1(X_1) & \longleftarrow & \tilde{\varphi}_1(X_1) \times \tilde{\varphi}_2(X_2) & \longrightarrow & \tilde{\varphi}_2(X_2) = \underline{A}^{\pi_2}(\tilde{\varphi}_2(X_2)) \end{array}$$

which is easily seen to be, in fact, an $\underline{A}^{H_1 \times H_2}$-map.

<u>Lemma 3.6.</u> Let $\varphi_i\colon G_i \to H_i$ and X_i be as in Definition 3.3 ($i = 1,2$). If X_i is a coflat Galois G_i-object, then the $\underline{A}^{H_1 \times H_2}$-map
$L_{X_1,X_2}\colon (\widetilde{\varphi_1 \times \varphi_2})(X_1 \times X_2) \to \tilde{\varphi}_1(X_1) \times \tilde{\varphi}_2(X_2)$ is an isomorphism.

Proof. By Theorem 2.20 and Proposition 3.2, $(\widetilde{\varphi_1 \times \varphi_2})(X_1 \times X_2)$ and $\tilde{\varphi}_1(X_1) \times \tilde{\varphi}_2(X_2)$ are coflat Galois $H_1 \times H_2$-objects, whence L_{X_1,X_2} is an isomorphism by Theorem 1.12.

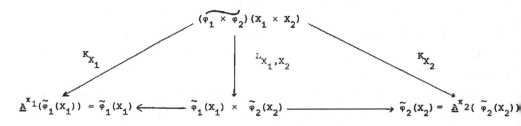

<u>Definition 3.7.</u> Given G_i in $|Gp(\underline{A})|$ $(i = 1,2)$, define pointed set maps $P_{G_1,G_2}: E(G_1) \times E(G_2) \to E(G_1 \times G_2)$ and $Q_{G_1,G_2}: E(G_1 \times G_2) \to E(G_1) \times E(G_2)$ as follows

$$P_{G_1,G_2}(cl(X_1), cl(X_2)) = cl(X_1 \times X_2)$$

$$Q_{G_1,G_2}(cl(Y)) = (cl(\tilde{\nu}_1(Y)), cl(\tilde{\nu}_2(Y))) = (E(\nu_1)(cl(Y)), E(\nu_2)(cl(Y))$$

P_{G_1,G_2} is well-defined by Proposition 3.2 and 2.7, and obviously preserves the base-point. The properties of E, discussed in Definition 2.22, guarantee that Q_{G_1,G_2} is likewise well-defined and preserves base-point.

<u>Proposition 3.8.</u> (a) If $\varphi_i: G_i \to H_i$ in $Gp(\underline{A})$, then the diagram below commutes

$$\begin{array}{ccc}
E(G_1) \times E(G_2) & \xrightarrow{P_{G_1,G_2}} & E(G_1 \times G_2) \\
\Big\downarrow {\scriptstyle E(\varphi_1) \times E(\varphi_2)} & & \Big\downarrow {\scriptstyle E(\varphi_1 \times \varphi_2)} \\
E(H_1) \times E(H_2) & \xrightarrow{P_{H_1,H_2}} & E(H_1 \times H_2)
\end{array}$$

(b) If G_i are in $|Gp(\underline{A})|$ $(i = 1,2)$, then P_{G_1,G_2} is an isomorphism, with $Q_{G_1,G_2} = P^{-1}_{G_1,G_2}$.

Proof. (a) Let X_i be coflat Galois G_i-objects $(i = 1,2)$. An easy computation then gives $E(\varphi_1 \times \varphi_2)\{P_{G_1,G_2}(cl(X_1), cl(X_2))\} = cl\{(\widetilde{\varphi_1 \times \varphi_1})(X_1 \times X_2)\}$ and $P_{H_1,H_2}\{E(\varphi_1) \times E(\varphi_2)(cl(X_1), cl(X_2))\} = cl\{\tilde{\varphi}_1(X_1) \times \tilde{\varphi}_2(X_2)\}$. (a) then follows from Lemma 3.6.

(b) With X_i as in (a), we have that $Q_{G_1,G_2}\{P_{G_1,G_2}(cl(X_1), cl(X_2))\} = (cl(\tilde{\nu}_1(X)), cl(\tilde{\nu}_2(X)))$ with $X = X_1 \times X_2$ and $\nu_i: G_1 \times G_2 \to G_i$ the projections. Now, the projection map $p_i: X \to \underline{A}^{\nu_i}(X_i)$, an $\underline{A}^{G_1 \times G_2}$-map, gives rise by adjointness to an $\underline{A}^{G_i}$- map $\tilde{\nu}_i(X) \to X_i$ which is an isomorphism, by Theorem 1.12. It then follows that $Q_{G_1,G_2}P_{G_1,G_2} = 1_{E(G_1 \times G_2)}$.

Conversely, let Y be a coflat Galois $G_1 \times G_2$-object. Then $P_{G_1,G_2}\{Q_{G_1,G_2}(cl(Y))\} = cl(\tilde{\nu}_1(Y) \times \tilde{\nu}_2(Y))$. Now, the adjointness transformations

θ^i: $1_{\underline{A}^{G_1 \times G_2}} \to \underline{A}^{\nu_i}\tilde{\nu}_i$ give $\underline{A}^{G_1 \times G_2}$-maps θ^i_Y: $Y \to \underline{A}^{\nu_i}\tilde{\nu}_i(Y)$, from which we obtain an $\underline{A}$-map $Y \to \tilde{\nu}_1(Y) \times \tilde{\nu}_2(Y)$ which is, in fact, easily seen to be an $\underline{A}^{G_1 \times G_2}$-map. It is the an isomorphism, by Theorem 1.12. We may then conclude that $P_{G_1,G_2}Q_{G_1,G_2} = 1_{E(G_1) \times E(G_2)}$ and the proof is complete.

<u>Theorem 3.9.</u> (a) If G is in $|Ab(\underline{A})|$, then E(G) is an abelian group, with addition defined to be the composition

$$E(G) \times E(G) \xrightarrow{P_{G_1,G_2}} E(G \times G) \xrightarrow{E(\mu)} E(G)$$

with μ: $G \times G \to G$ the multiplication of G. That is, $cl(X) + cl(Y) = cl(\tilde{\mu}(X \times Y))$. cl(G) is the zero element of E(G).

(b) If φ: $G \to H$ in $Ab(\underline{A})$, then $E(\varphi)$: $E(G) \to E(H)$ is a homomorphism.

Proof. Let X_i $(i = 1,2,3)$ be coflat Galois G-objects. Then, by Lemma 3,6, $\tilde{\mu}(X_1 \times X_2) \times X_3 \approx (\widetilde{\mu \times G})(X)$ in $\underline{A}^{G \times G}$, whence $\tilde{\mu}\{\tilde{\mu}(X_1 \times X_2) \times X_3\} \approx \tilde{\mu}(\widetilde{\mu \times G})(X) \approx \widetilde{\mu(\mu \times G)}($ in $\underline{A}^G$, with $X = X_1 \times X_2 \times X_3$. By a similar argument, $\tilde{\mu}\{X_1 \times \tilde{\mu}(X_2 \times X_3)\} \approx \widetilde{\mu(G \times \mu)}(X)$. The associativity of multiplication in G then gives the isomorphism

$$\tilde{\mu}\{\tilde{\mu}(X_1 \times X_2) \times X_3\} \approx \tilde{\mu}\{X_1 \times \tilde{\mu}(X_2 \times X_3)\} ,$$

from which the equality

$$(cl(X_1) + cl(X_2)) + cl(X_3) = cl(X_1) + (cl(X_2) + cl(X_3))$$

immediately follows. Thus addition in E(G) is associative. We delay for a moment the proof that it is commutative.

The proofs that E(G) possesses a zero element and additive inverses are similar; to avoid repetition we exhibit only the latter argument. Let η: $1 \to G$ and λ: $G \to G$ be the identity and inverse of G, respectively, and scrutinize the diagram

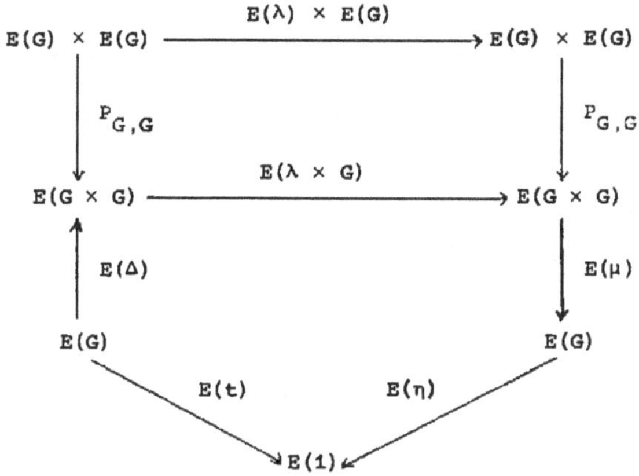

where Δ: $G \to G\times G$ is the diagonal map and t: $G \to 1$. The square commutes by Proposition 3.8, and the lower trapezoid commutes by properties of λ and the fact that E is a functor. Let ν_2: $G \times G \to G$ be the projection on the second factor; then $\nu_2\Delta = 1_G$ gives $E(\nu_2)E(\Delta) = 1_{E(G)}$, and thus, if X is a coflat Galois G-object,

$$Q_{G,G}\{E(\Delta)(cl(X))\} = (cl(Y),cl(X))$$

for some coflat Galois G-object Y. Since $P_{G,G}Q_{G,G} = 1_{E(G \times G)}$ by Proposition 3.8, tracing through the above diagram gives easily that

$$cl(Y) + cl(X) = E(\eta)(cl(1)) = cl(G) .$$

It follows that E(G) possesses left inverses, and is thus a group.

(b) Given φ: $G \to H$ in $|Ab(\underline{A})|$, we have the diagram

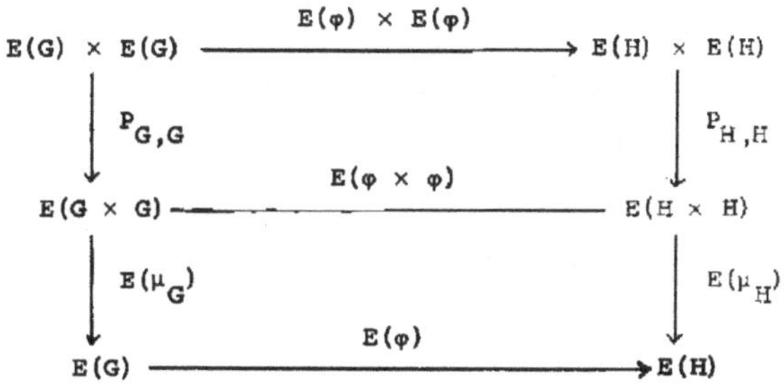

which is commutative by Proposition 3.8 and the functoriality of E. (b) follows immediately.

Finally, (b) guarantees that $Q_{G_1,G_2}: E(G_1 \times G_2) \to E(G_1) \times E(G_2)$ is a homomorphism, and thus $P_{G_1,G_2} = Q^{-1}_{G_1,G_2}$ is, too. It then follows from (b) that the addition map $+: E(G) \times E(G) \to E(G)$ is a homomorphism of groups, from which we obtain easily that E(G is abelian.

<u>Definition and Remark 3.10.</u> If J is in $|Ab(\underline{A})|$, we shall denote by $\mathfrak{X}(J)$ the abelian group constructed above from E(J). We then obtain a functor $\mathfrak{X}: Ab(\underline{A}) \to Ab$ which renders the diagram below commutative

$$\begin{array}{ccc} Ab(\underline{A}) & \xrightarrow{\;i\;} & Gp(\underline{A}) \\ \downarrow{\scriptstyle \mathfrak{X}} & & \downarrow{\scriptstyle E} \\ Ab & \xrightarrow{\;u\;} & \text{Pointed Sets} \end{array}$$

with i and u the inclusion and forgetful functors, respectively.

<u>Theorem 3.11.</u> The functor $\mathfrak{X}: Ab(\underline{A}) \to Ab$ is additive.

Proof. The assertion follows from Proposition 3.8 and Theorem 3.9, via well-known and easy argument which we omit.

We end this section with a discussion of the behaviour of the groups $\mathfrak{X}(J)$ under th application of certain functors.

<u>Theorem 3.12.</u> Let $\underline{A}, \underline{B}$ be categories with finite products and coequalizers, and $V: \underline{A} \to \underline{B}$ be a functor which preserves finite products. Assume further that, if X is f.c in $\underline{A}$, then V(X) is f.c. in $\underline{B}$.

(a) If G is a coflat group in $\underline{A}$ and X is a coflat Galois G-object, then V(G) is a coflat group in $\underline{B}$ and V(X) is a coflat Galois V(G)-object.

(b) If $\varphi: G \to H$ is a homomorphism of coflat groups in $\underline{A}$ and X is a coflat Galois G-object, then $\widetilde{V(\varphi)}(V(X)) \approx V(\tilde{\varphi}(X))$ in $\underline{B}^{V(H)}$.

(c) If J and V(J) are in $|Ab(\underline{A})|$ and $|Ab(\underline{B})|$, respectively, then the mapping $\mathfrak{X}(V): \mathfrak{X}(J) \to \mathfrak{X}(V(J))$ defined by the formula

$$\mathfrak{X}(V)(cl(X)) = cl(V(X))$$

is a homomorphism.

(d) If $\varphi: J \to J'$ in $Ab(\underline{A})$ and $V(\varphi): V(J) \to V(J')$ in $Ab(\underline{B})$, then the diagram below commutes

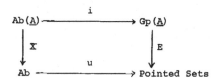

$$\begin{array}{ccc} \mathbb{X}(J) & \xrightarrow{\mathbb{X}(V)} & \mathbb{X}(V(J)) \\ \downarrow{\scriptstyle \mathbb{X}(\varphi)} & & \downarrow{\scriptstyle \mathbb{X}(V(\varphi))} \\ \mathbb{X}(J') & \xrightarrow{\mathbb{X}(V)} & \mathbb{X}(V(J')) \end{array}$$

Proof. (a) Our hypotheses on V guarantee, via Remarks 1.3 and 2.8, that $V(G)$ is a coflat group in $\underline{B}$, and $V(X)$ is in $|\underline{B}^{V(G)}|$ and is f.c. in $\underline{B}$. Furthermore, since X is a Galois G-object, $\gamma_{V(X)} = V(\gamma_X)$ is an isomorphism, the equality arising from the identifications of Remark 1.3. We may then conclude that $V(X)$ is a coflat Galois $V(G)$-object.

(b) Since V preserves products, we obtain easily that the diagram below commutes up to equivalence

$$\begin{array}{ccc} \underline{A}^H & \xrightarrow{\underline{A}^{\varphi}} & \underline{A}^G \\ \downarrow{\scriptstyle V} & & \downarrow{\scriptstyle V} \\ \underline{B}^{V(H)} & \xrightarrow{\underline{B}^{V(\varphi)}} & \underline{B}^{V(G)} \end{array}$$

Applying V to the adjointness $\underline{A}^G$-map $\theta_X\colon X \to \underline{A}^{\varphi}\tilde{\varphi}(X)$ then gives a $\underline{B}^{V(G)}$-map

$$V(X) \xrightarrow{V(\theta_X)} V\{\underline{A}^{\varphi}\tilde{\varphi}(X)\} \xrightarrow{\approx} \underline{B}^{V(\varphi)}(V(\tilde{\varphi}(X)))$$

from which we obtain by adjointness a $\underline{B}^{V(H)}$-map

(3.13) $$\widetilde{V(\varphi)}(V(X)) \longrightarrow V(\tilde{\varphi}(X))$$

(a) and Theorem 2.20 guarantee that both the domain and range of this map are Galois $V(H)$-objects, and thus the map is an isomorphism, by Theorem 1.12.

(c) Observe first that $\mathbb{X}(V)$ is well-defined, by (a). (b) guarantees easily that V preserves all of the data used to define the group structure on $\mathbb{X}(J)$, and thus $\mathbb{X}(V)$ is a homomorphism.

(d) This is an immediate consequence of (b).

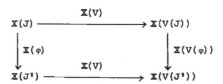

4. Galois Algebras and Coalgebras

Throughout this section R will be a fixed commutative ring with unit, and unadorn $\otimes$ will mean $\otimes_R$. We shall discuss Galois objects in the categories $\underline{A}$ and $\underline{C}$ of commutati R-algebras and cocommutative coalgebras, respectively, and homomorphisms of such. We refer the reader to [27, pp. 197-198] for the definitions of coalgebras and Hopf algebras, with the warning that the grading appearing there should be ignored.

Given C in $\underline{C}$, we shall denote the comultiplication (i.e., diagonal map) and couni (i.e., augmentation) of C by $\Delta_C\colon C \to C \otimes C$ and $\varepsilon_C\colon C \to R$, respectively. Note that R is an object of $\underline{C}$, with $\Delta_R = 1_R = \varepsilon_R\colon R \to R \otimes R = R$; furthermore, if C is in $|\underline{C}|$, then $\varepsilon_C\colon C \to R$ is the unique $\underline{C}$-map from C to R, and thus R is a terminal object of $\underline{C}$.

Given X,Y in $|\underline{C}|$, we give to $X \otimes Y$ a coalgebra structure as follows. The comultiplication $\Delta_{X \otimes Y}\colon X \otimes Y \to (X \otimes Y) \otimes (X \otimes Y)$ is defined to be the composition

$$X \otimes Y \xrightarrow{\Delta_X \otimes \Delta_Y} (X \otimes X) \otimes (Y \otimes Y) \longrightarrow (X \otimes Y) \otimes (X \otimes Y)$$

the unlabeled map being the so-called "middle four interchange" [27 ,p. 194]. The couni of $X \otimes Y$ is $\varepsilon_{X \otimes Y} = \varepsilon_X \otimes \varepsilon_Y\colon X \otimes Y \to R \otimes R = R$. It is easily verified that $X \otimes Y$ is i $|\underline{C}|$. In addition, if C is in $|\underline{C}|$, then the cocommutativity of C ensures that $\Delta_C\colon C \to C \otimes C$ is a $\underline{C}$-map.

The properties of $\underline{C}$ with regard to products and coequalizers are set forth in the proposition below.

Proposition 4.1. (a) If X,Y are in $|\underline{C}|$, then we have the product diagram

$$X \xleftarrow{\;i\;} X \otimes Y \xrightarrow{\;j\;} Y$$

with $i(x \otimes y) = x\varepsilon_Y(y)$ and $j(x \otimes y) = \varepsilon_X(x)y$. If $f\colon C \to X$, $g\colon C \to Y$ in $\underline{C}$, then the uni $\underline{C}$-map $h\colon C \to X \otimes Y$ such that $ih = f$, $jh = g$ is the composition

$$C \xrightarrow{\Delta_C} C \otimes C \xrightarrow{f \otimes g} X \otimes Y$$

(b) If $f,g: X \to Y$ in $\underline{C}$, then we have the following c.d. in $\underline{C}$

$$X \underset{g}{\overset{f}{\rightrightarrows}} Y \xrightarrow{\;h\;} C$$

where $C = \operatorname{Coker}(f - g)$, h is the canonical surjection, and $\Delta_C: C \to C \otimes C$ and $\varepsilon_C: C \to R$ are uniquely determined by the requirement that h be a $\underline{C}$-map.

Proof. Both parts of this proposition are well-known and easily verified (see, e.g., [28]). We shall check only that, if C is chosen as in (b), then Δ_C is well-defined. Clearly this amounts to showing that, if $I = \operatorname{Im}(f - g)$, then

$$\Delta_Y(I) \subseteq \operatorname{Im}\{(I \otimes Y) \oplus (Y \otimes I) \to Y \otimes Y\} .$$

If x is in X, write $\Delta_X(x) = \sum_{(x)} x_{(1)} \otimes x_{(2)}$ (this notational device will be discussed more systematically in Section 7). Then, since $f,g: X \to Y$ are $\underline{C}$-maps, we have that

$$\Delta_Y(f(x) - g(x)) = (f \otimes f - g \otimes g)(\Delta_X(x)) = \sum_{(x)} f(x_{(1)}) \otimes f(x_{(2)}) - \sum_{(x)} g(x_{(1)}) \otimes g(x_{(2)}) =$$

$$= \sum_{(x)} [f(x_{(1)}) - g(x_{(1)})] \otimes f(x_{(2)}) + \sum_{(x)} g(x_{(1)}) \otimes [f(x_{(2)}) - g(x_{(2)})].$$

The desired inclusion follows.

Corollary 4.2. Every object of $\underline{C}$ is coflat.

Proof. This follows from Proposition 4.1 and the fact that the functor $X \otimes (\ )$ is right exact on the category of R-modules.

Before turning to R-algebras, we interpret for coalgebras some of the definitions and constructions of the preceding sections. The statements below are easily verified.

Remarks 4.3. Let H be a group in $\underline{C}$.

(a) H is a cocommutative Hopf R-algebra [27, p. 198], trivially graded and with antipode [26]. The coalgebra operations in H are those which it possesses in virtue of the fact that it is an object of $\underline{C}$. The algebra multiplication on H is induced by the group multiplication $\mu_H: H \otimes H \to H$. $\eta_H(1)$ is the multiplicative identity of the R-algebra H.

(b) If C is in $|\underline{C}^H|$, then C is a right H-module via the formula

(4.4) $$cu = \alpha_C(c \otimes u) \qquad (c \text{ in } C,\ u \text{ in } H)$$

Furthermore, since Δ_C is a $\underline{C}$-map, the following identity holds

(4.5) $$\Delta_C(cu) = \Delta_C(c)\Delta_H(u) \qquad (c \text{ in } C,\ u \text{ in } H)$$

the right $H \otimes H$ -module structure on $C \otimes C$ arising in the obvious way from the right H-module structure on C. In other words, C is a "right H-module coalgebra" in sense of Sweedler [42].

On the other hand, if C in $|\underline{C}|$ is a right H-module satisfying (4.5), then C is an object of $\underline{C}^H$ if α_C is defined by (4.4). These observations establish an isomorphism of $\underline{C}^H$ with the category of which the objects are all objects of $\underline{C}$ which are simultaneously right H-modules satisfying (4.5), and the maps are all $\underline{C}$-maps which are simultaneously H-module homomorphisms. We shall view this isomorphism as an identification.

(c) If C is in $|\underline{C}^H|$, then the map $\gamma_C\colon C \otimes H \to C \otimes C$ of (1.1) satisfies the formula

$$\gamma_C(c \otimes u) = \Delta_C(c)(1 \otimes u) \qquad (c \text{ in } C,\ u \text{ in } H)$$

(d) Let $\varphi\colon H \to H'$ be a homomorphism of groups in $\underline{C}$, and C be an H-object. View H' as a left H-module via φ. Then $C' = C \otimes_H H'$ is an H'-object, the right H'-module structur being the obvious one and the coalgebra operations on C' satisfying the formulae

$$\Delta_{C'}(c \otimes u') = \sum_{(c),(u')} (c_{(1)} \otimes u'_{(1)}) \otimes (c_{(2)} \otimes u'_{(2)})$$

$$\varepsilon_{C'}(c \otimes u') = \varepsilon_C(c)\varepsilon_{H'}(u')$$

where $\Delta_C(c) = \sum_{(c)} c_{(1)} \otimes c_{(2)}$, $\Delta_{H'}(u') = \sum_{(u')} u'_{(1)} \otimes u'_{(2)}$ for c in C, u' in H'. Furthermore the diagram

$$C \otimes H \otimes H' \underset{\alpha_C \otimes H'}{\overset{\omega_{C,\varphi}}{\rightrightarrows}} C \otimes H' \xrightarrow{\ \pi_C\ } C'$$

is a c.d. in $\underline{C}$, where $\omega_{C,\varphi}(c \otimes u \otimes u') = c \otimes \varphi(u)u'$ as in (2.10) and π_C is the canonical map. Thus, by Theorem 2.9, there is a $\underline{C}^H$-isomorphism $\tilde{\varphi}(C) \overset{\approx}{\to} C'$ rendering the diagram below commutative

$$\begin{array}{ccc} & & \tilde{\varphi}(C) \\ & \overset{\rho_{C,\varphi}}{\nearrow} & \\ C \otimes H' & & \downarrow \approx \\ & \underset{\pi_C}{\searrow} & \\ & & C' \end{array}$$

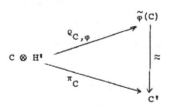

That is, $C' \approx \tilde{\varphi}(C)$ in $\underline{C}^H$.

Galois objects in the category $\underline{C}$ will reappear later in this section, and will play a role in Chapter III. But for the moment we turn to a consideration of the category $\underline{A}$ of commutative R-algebras. We would like to apply the theory developed in previous sections to $\underline{A}^{op}$. However, we prefer to express our results in terms of the category $\underline{A}$; this preference will, as usual, necessitate the constant use of the prefix "co-". Thus we shall discuss cogroups in $\underline{A}$, coproducts, etc., the definitions of which are obtained by dualization from the definitions of groups in a category, products, etc. Note that the role of the categories $(\underline{A},X)$ of Section 1 will now be played by the categories $(S,\underline{A})$ (S in $|\underline{A}|$). Of course, $(S,\underline{A})$ is simply the category of commutative S-algebras.

We shall depart from the convention just set forth in that we shall say "flat", and "equalizer" rather than "co-coflat", etc. Finally, we shall not use the term "co-object".

Now, it is well-known and easy to see that $\underline{A}$ possesses finite colimits. In particular, the coproduct in $\underline{A}$ arises from $\otimes$. That is, if A,B are in $|\underline{A}|$, then

$$A \xrightarrow{\ i\ } A \otimes B \xleftarrow{\ i\ } B$$

is a coproduct diagram in $\underline{A}$, with $i(a) = a \otimes 1$, $j(b) = 1 \otimes b$. Furthermore, if $f,g\colon B \to C$ in $\underline{A}$, then

$$A \xrightarrow{\ h\ } B \underset{g}{\overset{f}{\rightrightarrows}} C$$

is an equalizer diagram in $\underline{A}$, with $A = \{b \text{ in } B/\ f(b) = g(b)\}$ and h the inclusion. Thus $\underline{A}$ satisfies the hypotheses dual to those set forth at the beginning of Section 2. Finally, S in $|\underline{A}|$ is a flat object of $\underline{A}$ if and only if the functor $S \otimes (\ )\colon \underline{A} \to \underline{A}$ preserves equalizers.

<u>Proposition 4.6.</u> Let S be in $|\underline{A}|$.

(a) S is a cofaithful object of $\underline{A}$ if and only if the following condition holds: Whenever $f\colon M \to N$ is a homomorphism of R-modules such that $S \otimes f\colon S \otimes M \to S \otimes N$ is an isomorphism, then f is likewise an isomorphism.

(b) S is a flat object of $\underline{A}$ if and only if S is a flat R-module in the sense of [10 , p. 122].

(c) S is cofaithful and flat as an object of $\underline{A}$ if and only if S is a faithfully flat R-module in the sense of [10 , Chapter I, §3].

Proof. Let $\underline{M}_R$ be the category of R-modules. If M is in $|\underline{M}_R|$, we shall denote by $M_R = M \oplus R$ the "split singular extension" of R by M, an object of $\underline{A}$ [11, pp. 293-295]. The multiplication in M_R satisfies the formula

$$(m_1,r_1)(m_2,r_2) = (m_1r_2 + m_2r_1,r_1r_2)$$

If f: $M \to N$ in $\underline{M}_R$, we obtain f_R: $M_R \to N_R$ in $\underline{A}$ defined by $f_R(m,r) = (f(m),r)$. It is easy to see that we thus obtain a functor $(\ )_R$: $\underline{M}_R \to \underline{A}$. Furthermore, if

$$K \xrightarrow{\ h\ } M \overset{f}{\underset{g}{\rightrightarrows}} N$$

is an equalizer diagram in $\underline{M}_R$ (i.e., the sequence

$$0 \longrightarrow K \xrightarrow{\ h\ } M \xrightarrow{\ f-g\ } N$$

is exact), then a routine computation establishes that

$$K_R \xrightarrow{\ h_R\ } M_R \overset{f_R}{\underset{g_R}{\rightrightarrows}} N_R$$

is an equalizer diagram in $\underline{A}$, and conversely.

(a) Suppose that S is a cofaithful object of $\underline{A}$. Let f: $M \to N$ be a homomorphism of R-modules such that $S \otimes f$: $S \otimes M \to S \otimes N$ is an isomorphism. Then f_R: $M_R \to N_R$ is easily seen to be a homomorphism of R-algebras such that $S \otimes f_R$: $S \otimes M_R \to S \otimes N_R$ is an isomorphism, whence f_R is likewise an isomorphism. But then f is an isomorphism, and the condition set forth in (a) holds. The converse is trivial.

(b) Suppose that S is a flat object of $\underline{A}$. If

$$0 \longrightarrow K \xrightarrow{\ h\ } M \xrightarrow{\ f\ } N$$

is an exact sequence in $\underline{M}_R$, then

$$K \xrightarrow{\ h\ } M \overset{f}{\underset{0}{\rightrightarrows}} N$$

is an equalizer diagram in $\underline{M}_R$, whence

$$K_R \xrightarrow{\ h_R\ } M_R \overset{f_R}{\underset{0_R}{\rightrightarrows}} N_R$$

is an equalizer diagram in $\underline{A}$, by the above remarks. But then

$$S \otimes K_R \xrightarrow{\;S \otimes h_R\;} S \otimes M_R \underset{S \otimes 0_R}{\overset{S \otimes f_R}{\rightrightarrows}} S \otimes N_R$$

is an equalizer diagram in $\underline{A}$; in particular, $S \otimes h_R$ is one-to-one. But the commutative diagram

$$\begin{array}{ccc} S \otimes K & \xrightarrow{\;S \otimes h\;} & S \otimes M \\ \downarrow & & \downarrow \\ S \otimes K_R & \xrightarrow{\;S \otimes h_R\;} & S \otimes M_R \end{array}$$

then shows that $S \otimes h$ is one-to-one (the unlabeled maps here arise from the split monomorphisms $K \to K_R$, $M \to M_R$, hence are one-to-one). We have shown that $S \otimes (\ )$ preserves monomorphisms of R-modules, from which it follows that S is a flat R-module. The converse part of (b) follows trivially from the definitions; we shall omit the details.

(c) is then an immediate consequence of (a) and (b), and the proof is complete.

In the proposition below we interpret, for the category $\underline{A}$ of commutative R-algebras, some of the material of the preceding sections.

<u>Proposition and Remarks 4.7.</u> (a) A cogroup A in the category $\underline{A}$ is a commutative Hopf algebra with antipode. The algebra operations on A are those arising from the fact that A is an object of $\underline{A}$. The comultiplication Δ_A: $A \to A \otimes A$, counit ε_A: $A \to R$, and antipode λ_A: $A \to A$ arise from the fact that A^{op} is a group in $\underline{A}^{op}$.

(b) An A-object in $\underline{A}$ is a pair $S = (S,\alpha_S)$, with S a commutative R-algebra and α_S: $S \to S \otimes A$ an R-algebra homomorphism such that

$$(\alpha_S \otimes A)\alpha_S = (S \otimes \Delta_A)\alpha_S \colon S \to S \otimes A \otimes A$$

and

$$(S \otimes \varepsilon_A)\alpha_S = 1_S \colon S \to S \otimes R = S\ .$$

(c) If S is as in (b), then the mapping γ_S: $S \otimes S \to S \otimes A$ of (1.1) satisfies the formula $\gamma_S(x \otimes y) = (x \otimes 1)\alpha_S(y)$ for x,y in S.

(d) Let φ: $A \to B$ be a homomorphism of cogroups in $\underline{A}$, and the functor $\underline{A}^{\varphi}$: $\underline{A}^A \to \underline{A}^B$ be as in Theorem 2.9. If S is an A-object, then $\underline{A}^{\varphi}(S) = S$, with $\alpha_{\underline{A}^{\varphi}(S)}$: $S \to S \otimes B$ the composite

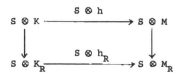

$$S \xrightarrow{\alpha_S} S \otimes A \xrightarrow{S \otimes \varphi} S \otimes B$$

(e) Let $\varphi: A \to B$ be a homomorphism of flat cogroups in $\underline{A}$, and S be in $\underline{A}^B$. If $\tilde{\varphi}: \underline{A}^B \to \underline{A}^A$ is as in Theorem 2.9, then we have the following equalizer diagram

$$\tilde{\varphi}(S) \xrightarrow{\varrho_{S,\varphi}} S \otimes A \underset{\alpha_S \otimes A}{\overset{\omega_{S,\varphi}}{\rightrightarrows}} S \otimes B \otimes A$$

where $\omega_{S,\varphi}$ is the composite

$$S \otimes A \xrightarrow{S \otimes \Delta_A} S \otimes A \otimes A \xrightarrow{S \otimes \varphi \otimes A} S \otimes B \otimes A$$

Thus $\tilde{\varphi}(S)$ is the subalgebra of $S \otimes A$ consisting of all elements which have the same image under $\omega_{S,\varphi}$ and $\alpha_S \otimes A$, and $\varrho_{S,\varphi}$ is the inclusion map.

(f) X: Cogp($\underline{A}$) $\to$ Ab is a contravariant functor on the category of flat cogroups in $\underline{A}$.

Proof. The above statements are all easy consequences of the preceding remarks and the general theory of Section 1-3; we omit the details.

Let us now consider duality. Let $\underline{A}_o$ be the full subcategory of $\underline{A}$, the objects of which are all objects of $\underline{A}$ which are finitely generated projective R-modules; define similarly the full subcategory $\underline{C}_o$ of $\underline{C}$. Given C in $|\underline{C}_o|$, observe that $C^* = \mathrm{Hom}_R(C,R)$ is in $|\underline{A}_o|$, the algebra operations on C^* being obtained from the coalgebra operations on C by applying the functor $(\ )^*$. For example, the multiplication on C^* is given by the map $\Delta_C^*: C^* \otimes C^* = (C \otimes C)^* \to C^*$. The equality here arises from viewing the natural isomorphism $A^* \otimes B^* \approx (A \otimes B)^*$ as an identification; this isomorphism always exists if A and B are finitely generated projective R-modules. Thus we obtain a functor $(\ )^*: \underline{C}_o^{op} \to \underline{A}_o$ which is easily seen to be an isomorphism of categories. The proposition below is an immediate consequence of this fact.

Proposition 4.8. If H is a group in $\underline{C}_o$, then H^* is a cogroup in $\underline{A}_o$, the operations on H^* arising via duality from their counterparts on H. If C is an H-object in $\underline{C}_o$, then C^* is in like manner an H^*-object in $\underline{A}_o$. Furthermore, C is a Galois H-object if and only if C^* is a Galois H^*-object. Finally, duality gives rise to an isomorphism $X_{\underline{C}_o}(H) \approx X_{\underline{A}_o}(H^*)$ which is natural in H.

Remarks 4.9. The preceding proposition raises naturally the following question: If H is a group in $\underline{C}$ which is a finitely generated projective R-module and C is a Galois H-object, then is C likewise a finitely generated projective R-module? We have no answer to this question, although it is true for the special case in which H = RG, the group algebra with coefficients in R of the finite group G. However, the analogous question for Galois objects in the category $\underline{A}$ has an affirmative answer; see Theorem 12.3.[2]

We end this section with several concrete examples of Galois objects in the category $\underline{A}$.

Example 4.10. Let G be a finite group, and GR = (RG)*, the cogroup in $\underline{A}$ which is the dual of the group algebra RG. Then the Galois GR-objects are precisely the "Galois extensions of R with Galois group G" in the sense of [5] and [12]. This example will be discussed more thoroughly in Section 7.

Example 4.11. Let k be a field of characteristic $p \neq 0$, and define the k-algebra A by $A = k[t]/(t^{p^n})$, n a positive integer. Write A = k[x], where $t \to x$ under the canonical map $k[t] \to A$. It is easily verified that the formulae below yield on A the structure of a commutative Hopf algebra with antipode, in virtue of which A is a cogroup in the category $\underline{A}$ of commutative k-algebras.

$$\Delta_A(x) = x \otimes 1 + 1 \otimes x$$
$$\varepsilon_A(x) = 0$$
$$\lambda_A(x) = -x$$

Now let $K = k(\omega)$ be a purely inseparable primitive extension field of k of exponent n; i.e., ω^{p^n} is in k but $\omega^{p^{n-1}}$ is not in k. It is not difficult to show that K is a Galois A-object in $\underline{A}$, with $\alpha_K: K \to K \otimes A$ defined by the formulae $\alpha_K(\omega) = \omega \otimes 1 + 1 \otimes x$.

Our third example will require somewhat more preparation. Let $Z_n = Z/nZ$, which we shall write multiplicatively with generator τ. We shall analyze the Galois RZ_n-objects in the category $\underline{A}$ of commutative R-algebras. Of course, the algebra structure on RZ_n is the usual one, and the comultiplication, counit, and antipode satisfy the formulae

$$\Delta_{RZ_n}(\sigma) = \sigma \otimes \sigma$$
$$\varepsilon_{RZ_n}(\sigma) = 1 \qquad (\sigma \text{ in } Z_n)$$
$$\lambda_{RZ_n}(\sigma) = \sigma^{-1}$$

Our first observation is that an RZ_n-object structure on a commutative R-algebra S is tantamount to a Z_n-grading on S, which we now define. If m is an integer, we denote by $\overline{m}$ the integer uniquely determined by the conditions $0 \leq \overline{m} < n$, $\overline{m} \equiv m \pmod{n}$.

__Definition 4.12.__ A Z_n-grading on S in $|\underline{A}|$ is an R-module direct sum decomposition $S = S_0 \oplus S_1 \oplus \ldots \oplus S_{n-1}$ such that $S_i S_j \subseteq S_{\overline{i+j}}$. We say that an algebra homomorphism $f: S \to T$ of Z_n-graded algebras (i.e., algebras equipped with Z_n-grading) is of degree zero if $f(S_i) \subseteq T_i$ for $0 \leq i < n$. We shall denote by $\underline{A}(Z_n)$ the category of commutative Z_n-graded R-algebras and algebra homomorphisms of degree zero.

__Proposition 4.13.__ (a) Let S be in $|\underline{A}^{RZ_n}|$. Then S is Z_n-graded, an element x in S being in S_i $(0 \leq i < n)$ if and only if $\alpha_S(x) = x \otimes \tau^i$.

(b) Conversely, if S is in $|\underline{A}(Z_n)|$, then S is an RZ_n-object with $\alpha_S: S \to S \otimes RZ_n$ defined by $\alpha_S(x) = x \otimes \tau^i$ for x in S_i.

(c) These observations give rise to a category isomorphism $\underline{A}^{RZ_n} \approx \underline{A}(Z_n)$.

Proof. (a) Let S be in $|\underline{A}^{RZ_n}|$. If x is in $\underline{A}$, write

$$\alpha_S(x) = \sum_{i=0}^{n-1} x_i \otimes \tau^i$$

in $S \otimes RZ_n$, with x_i in S. Now, we have from Proposition 4.7(b) that

$$(\alpha_S \otimes RZ_n)\alpha_S = (S \otimes \Delta_{RZ_n})\alpha_S: S \to S \otimes RZ_n \otimes RZ_n$$

and

$$(S \otimes \varepsilon_{RZ_n})\alpha_{RZ_n} = 1_S: S \to S .$$

The second equality yields immediately that $x = \sum_{i=0}^{n-1} x_i$. We then have from the first equality that

$$\sum_{i=0}^{n-1} \alpha_S(x_i) \otimes \tau^i = \sum_{i=0}^{n-1} x_i \otimes \tau^i \otimes \tau^i$$

whence $\alpha_S(x_i) = x_i \otimes \tau^i$ for $0 \leq i < n$; i.e., x_i is in S_i.

Suppose now that $x = \sum_{i=0}^{n-1} y_i$, with y_i in S_i. Then

$$\sum_{i=0}^{n-1} x_i \otimes \tau^i = \alpha_S(x) = \sum_{i=0}^{n-1} \alpha_S(y_i) = \sum_{i=0}^{n-1} y_i \otimes \tau^i$$

nd thus $x_i = y_i$. We have shown that $S = S_0 \oplus S_1 \oplus \ldots \oplus S_{n-1}$. If x,y are in S_i, S_j, respectively, then

$$\alpha_S(xy) = \alpha_S(x)\alpha_S(y) = (x \otimes \tau^i)(y \otimes \tau^j) = xy \otimes \tau^{\overline{i+j}}$$

nd so xy is in $S_{\overline{i+j}}$. It follows that S is a Z_n-graded R-algebra. Note that, if f: $S \to T$ n $\underline{A}^{RZ_n}$ and x is in S_i, then $\alpha_T(f(x)) = (f \otimes RZ_n)(\alpha_S(x)) = f(x) \otimes \tau^i$, whence f(x) is in T_i; therefore f is a homomorphism of degree zero.

This establishes (a). (b) is routine, and (c) follows easily from (a) and (b); we mit the details.

We shall identify the categories $\underline{A}^{RZ_n}$ and $\underline{A}(Z_n)$ via the isomorphism established in he proceding proposition.

Lemma 4.14. Let S be a Galois RZ_n-object. Then

(a) The unit map $R \to S$ is one-to-one and has image precisely S_0. Thus it induces an somorphism $S_0 \approx R$ (and we shall identify these algebras via this isomorphism.)

(b) The multiplication in S induces R-module isomorphisms $I^i \approx S_i$ for $0 < i < n$ nd β: $I^n \overset{\approx}{\to} S_0 = R$, with $I = S_1$ and I^i the i-fold tensor power of I. In particular, I is n invertible R-module [10, Chapter II, §4].

Proof. Letting γ_S: $S \otimes S \to S \otimes RZ_n$ be as in Proposition 4.7 (c), we have that $\gamma_S(x \otimes y) = xy \otimes \tau^i$ for x in S and y in S_i. Thus $\gamma_S(S \otimes S_i) \subseteq S \otimes \tau^i$. Since S is a Galois RZ_n-object, γ_S is an isomorphism, whence γ_S maps $S \otimes S_i$ isomorphically onto $S \otimes \tau^i$. In articular, setting $i = 0$, we obtain that the mapping θ: $S \otimes S_0 \to S$ defined by $\theta(x \otimes x_0) = xx_0$ is an isomorphism of R-algebras. Now, if η: $R \to S_0$ is the unit map, then e have the commutative diagram

$$\begin{array}{ccc} S \otimes R & \xrightarrow{S \otimes \eta} & S \otimes S_0 \\ & \searrow & \downarrow \theta \\ & & S \end{array}$$

he unlabeled map being the usual isomorphism. Thus $S \otimes \eta$ is an isomorphism, whence η is ikewise because S is cofaithful. This establishes (a).

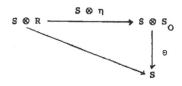

Turning now to (b), we denote by γ_i: $S \otimes S_i \overset{\approx}{\to} S \otimes \tau^i$ the restriction to $S \otimes S_i$ of he isomorphism γ_S: $S \otimes S \to S \otimes RZ_n$. If $0 < i,j < n$, then the multiplication in S yields an R-module homomorphism $\mu_{ij}: S_i \otimes S_j \to S_{\overline{i+j}}$. Since γ_S is a homomorphism of S-algebras, we then have the commutative diagram

$$\begin{array}{ccc}
S \otimes (S_i \otimes S_j) & \xrightarrow{\quad S \otimes \mu_{ij} \quad} & S \otimes S_{\overline{i+j}} \\
\| & & \| \\
(S \otimes S_i) \otimes_S (S \otimes S_j) & & S \otimes S_{\overline{i+j}} \\
\approx \Big\downarrow \gamma_i \otimes \gamma_j & & \approx \Big\downarrow \gamma_{\overline{i+j}} \\
(S \otimes T^i) \otimes_S (S \otimes T^i) & \xrightarrow{\quad \nu_{ij} \quad} & S \otimes T^{\overline{i+j}}
\end{array}$$

the unlabeled isomorphisms being the obvious ones, and ν_{ij} being the restriction of the multiplication map of $S \otimes RZ_n$. Since ν_{ij} is manifestly an isomorphism, and all vertical maps are likewise, it follows that $S \otimes \mu_{ij}$ is an isomorphism. Since S is cofaithful, we may apply Proposition 4.6 (a) to conclude that $\mu_{ij}: S_i \otimes S_j \to S_{\overline{i+j}}$ is an isomorphism. (b) then follows readily from (a), completing the proof of the lemma.

Lemma 4.14 assigns to a Galois RZ_n-object a pair (I,β), where I is an invertible R-module and $\beta: I^n \overset{\approx}{\to} R$ is an R-module isomorphism. Conversely, such a pair (I,β) gives rise to a Galois RZ_n-object $S = R(I,\beta)$ by means of a construction which we shall briefl sketch, omitting most details.

Define $S = R(I,\beta) = R \oplus I \oplus I^2 \oplus \dots \oplus I^{n-1}$ as an R-module, with multiplication given by the formulas below

$$(x_1 \otimes \dots \otimes x_i)(x_{i+1} \otimes \dots \otimes x_{i+j}) = \begin{cases} x_1 \otimes \dots \otimes x_{i+j} & \text{if } i+j < n \\ \beta(x_1 \otimes \dots \otimes x_n)(x_{n+1} \otimes \dots \otimes x_{i+j}) & \\ \text{if } i+j \geq n & \end{cases}$$

where x_k is in I and $0 < i,j < n$. That S is a well-defined commutative R-algebra follows readily from the lemma below.

<u>Lemma 4.15.</u> If I is an invertible R-module, then $y \otimes x = x \otimes y$ in $I \otimes I$ for any x,y in I. In particular, if $\beta: I^n \overset{\approx}{\to} R$ is an R-module isomorphism and $x_1,\dots,x_m$ is in I with $m > n$, then

$$\beta(x_1 \otimes \dots \otimes x_n)x_{n+1} \otimes \dots \otimes x_m = \beta(x_{\pi(1)} \otimes \dots \otimes x_{\pi(n)})x_{\pi(n+1)} \otimes \dots \otimes x_{\pi(m)}$$

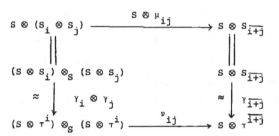

for any permutation π of the set of indices $\{1,\ldots,m\}$.

Proof. These statements are trivially true for the special case in which R is a local ring, for then $I \approx R$. The general case follows from a routine localization argument.

We define a Z_n-grading on $S = R \oplus I \oplus \ldots \oplus I^{n-1}$ by the condition that $S_i = I^i$ for $0 \leq i < n$. It is then easily verified that S is a Galois RZ_n-object. We summarize the preceding remarks in the example below.

<u>Example 4.16.</u> Lemma 4.14 and subsequent comments yield a one-to-one correspondence between Galois RZ_n-objects and pairs (I,β), where I is an invertible R-module and $\beta: I^n \overset{\approx}{\to} R$ is an R-module isomorphism. If (J,ζ) is another such pair, then the Galois RZ_n-objects corresponding to these pairs are isomorphic in $\underline{A}^{RZ_n}$ if and only if there is an R-module isomorphism $\kappa: I \overset{\approx}{\to} J$ rendering the diagram below commutative

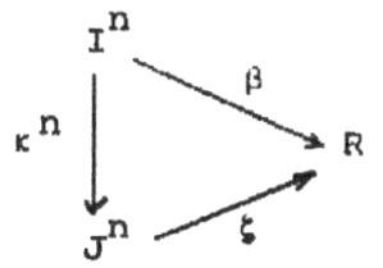

where $\kappa^n(x_1 \otimes \ldots \otimes x_n) = \kappa(x_1) \otimes \ldots \otimes \kappa(x_n)$ for x_i in I.

All statements of this example have been discussed above except for the last, which follows from a routine computation which we omit.

The proposition below, which will play a role in Chapter III, can be proved by a method quite similar to that used to establish Proposition 4.13. Let Z denote the additive group of rational integers, which we shall write multiplicatively with generator t.

<u>Proposition 4.17.</u> (a) Let S be in $|\underline{A}^{RZ}|$. Then S is Z-graded, an element x in S being in S_n if and only if $\alpha_S(x) = x \otimes t^n$ $(-\infty < n < +\infty)$.

(b) Conversely, if

$$S = \sum_{n=-\infty}^{+\infty} \oplus S_n$$

is a Z-graded commutative R-algebra, then S is an RZ-object with $\alpha_S: S \to S \otimes RZ$ defined by $\alpha_S(x) = x \otimes t^n$ for x in S_n.

(c) These observations give rise to a category isomorphism $\underline{A}^{RZ} \approx \underline{A}(Z)$, the latter being the category of commutative Z-graded R-algebras and R-algebra homomorphisms of degree zero.

Proposition 4.17 yields, via considerations similar to those of Example 4.16, a one-to-one correspondence between the Galois RZ-objects and the invertible R-modules. This correspondence is also set forth (in the language of principal homogeneous spaces) in [41 , Exposé 8, 2.31], with the additional requirement that the Galois RZ-objects considered be faithfully flat R-modules.

5. Triples, Extensions, and Cohomology

In this section we shall discuss the behaviour of Galois objects under the application of functors $U: \underline{A} \to \underline{B}$ which are "tripleable" in the sense of [7] and [9]. We shall have use for the triple cohomology groups with coefficients in an abelian group J in $\underline{A}$; these groups are defined in [7], and are denoted by $H^n(1,J)_1$, where 1 is, as usual, the terminal object of $\underline{A}$. (Actually, in the papers just cited, more general cohomology groups $H^n(W,J)_X$ are defined for an arbitrary X in $|\underline{A}|$ and W in $|(\underline{A},X)|$; however, we shall need only the special case $W = X = 1$). We shall write $H^n(U,J) = H^n(1,J)_1$. The group of n-cocycles, with coefficients in J, will be denoted by $Z^n(U,J)$. If $\varphi: J \to J'$ is a homomorphism of abelian groups in $\underline{A}$, the induced homomorphisms will be written as

$$Z^n(U,\varphi): Z^n(U,J) \to Z^n(U,J')$$

and

$$H^n(U,\varphi): H^n(U,J) \to H^n(U,J').$$

Throughout this section we shall assume, unless explicitly stated otherwise, that the categories with which we deal possess finite products and coequalizers. Note finally that, if U is tripleable, then it possesses a left adjoint [7] and thus preserves limits.

Lemma 5.1. Let $U: \underline{A} \to \underline{B}$ be tripleable and X be in $|\underline{A}|$. If $\underline{B}(1,U(X)) \neq \emptyset$, then X is faithful in $\underline{A}$.

Proof. Let $f: Y \to Y'$ in $\underline{A}$ be such that $X \times f: X \times Y \to X \times Y'$ is an isomorphism. Applying U, we obtain the commutative diagram

$$\begin{array}{ccc} U(X \times Y) & \xrightarrow{U(X \times f)} & U(X \times Y') \\ \| & & \| \\ U(X) \times U(Y) & \xrightarrow{U(X) \times U(f)} & U(X) \times U(Y') \end{array}$$

the vertical isomorphisms arising from the fact that U preserves products. Thus $U(X) \times U(f)$ is a $\underline{B}$-isomorphism. Our hypotheses on X guarantee, via Corollary 0.3, that $U(X)$ is faithful in $\underline{B}$, whence $U(f)$ is an isomorphism. But U, being tripleable, reflects isomorphisms; therefore f is an $\underline{A}$-isomorphism, and the proof is complete.

Definition 5.2. [9, Definition 4]. Let $U: \underline{A} \to \underline{B}$ be tripleable, and G be a group in $\underline{A}$. X in $|\underline{A}^G|$ will be called a principal G-object if the following conditions hold

(a) $\underline{B}(1, U(X)) \neq \emptyset$

(b) Given any f_1, f_2: $A \to X$ in $\underline{A}$, there exists a unique $\underline{A}$-map $g: A \to G$ such that the diagram below commutes

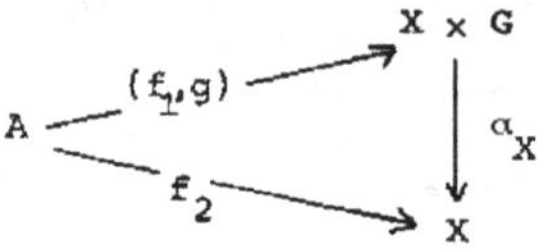

Proposition 5.3. Let $U: \underline{A} \to \underline{B}$ and G be as in Definition 5.2. Then X in $|\underline{A}^G|$ is a principal G-object if and only if $U(X) \approx U(G)$ in $\underline{B}^{U(G)}$. If these conditions hold, then X is a Galois G-object.

Proof. Note first that the statements of the Proposition make sense in virtue of Remark 1.3 and the fact that U preserves products. Suppose now that X is a principal G-object. Then X is faithful in $\underline{A}$, by Lemma 5.1. Furthermore, it is easy to see that (5.2b) is equivalent to condition (c) of Theorem 1.8. Thus, by theorem, X is a Galois G-object. Since U(X) is faithful in $\underline{B}$, by Corollary 0.3, we may apply Remark 1.3 (c) to obtain that U(X) is a Galois U(G)-object. That $U(X) \approx U(G)$ in $\underline{B}^{U(G)}$ then follows from Corollary 1.14.

Conversely, let X in $|\underline{A}^G|$ be such that $U(X) \approx U(G)$ in $|\underline{B}^{U(G)}|$; then clearly $\underline{B}(1, U(X)) \neq \emptyset$. We also have the commutative diagram

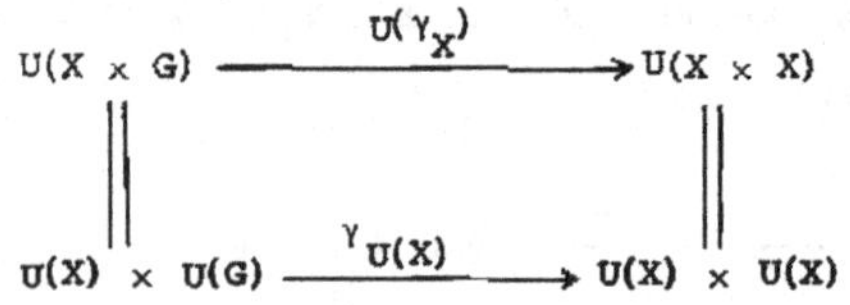

the vertical isomorphisms arising from Remark 1.3 and the fact that U, being tripleable, preserves products. Our hypotheses on X guarantee that $\gamma_{U(X)}$ is an isomorphism, whence $U(\gamma_X)$ is, too. Since U, being tripleable, reflects isomorphisms, we may conclude that γ_X is an isomorphism in $\underline{A}$. Thus, by Theorem 1.8, X satisfies (5.2 b), and is therefore a principal G-object. This completes the proof.

Theorem 5.4. Let $U: \underline{A} \to \underline{B}$ be tripleable, and assume that an object X of $\underline{A}$ is f.c. in $\underline{A}$ if and only if U(X) is f.c. in $\underline{B}$.

(a) There exists a natural transformation $\kappa: H'(U,-) \to \mathfrak{X}$ of functors from $Ab(\underline{A})$ to Ab. $\kappa_J: H'(U,J) \to \mathfrak{X}(J)$ is a monomorphism for each J in $|Ab(\underline{A})|$.

(b) If J is an abelian group in $\underline{A}$ such that U(J) is in $|Ab(\underline{B})|$, then J is in $|Ab(\underline{A})|$ and the sequence below is exact-

$$0 \longrightarrow H^1(U,J) \xrightarrow{\kappa_J} \mathfrak{X}(J) \xrightarrow{\mathfrak{X}(U)} \mathfrak{X}(U(J))$$

with $\mathfrak{X}(U)(cl(X)) = cl(U(X))$ as in Theorem 3.12.

Proof. Since U is tripleable, we may assume without loss of generality that $\underline{A} = \underline{B}^{\mathbf{T}}$, the category of algebras over a triple $\mathbf{T} = (T,\mu,\eta)$ in $\underline{B}$ in the sense of [7]. Here $T: \underline{B} \to \underline{B}$ is a functor and $\eta: 1_{\underline{B}} \to T$, $\mu: T^2 \to T$ are natural transformations satisfying the axioms of [7,p. 337]. An object of $\underline{A}$ is then a pair $X = (X,\xi_X)$, with X in $|\underline{B}|$ and $\xi_X: T(X) \to X$ a $\underline{B}$-map, the "triple structure map", satisfying the conditions of [7, p. 337]. An $\underline{A}$-map $f: X \to Y$ is then a $\underline{B}$-map $f: X \to Y$ such that $\xi_Y f = \xi_X$, and the functor $U: \underline{A} \to \underline{B}$ is the "underlying" functor which sends the pair (X,ξ_X) to the object X of $\underline{B}$. In the following discussion we shall often denote an object $X = (X,\xi_X)$ of $\underline{A}$ and its image in $\underline{B}$ by the same letter X, suppressing both the ξ_X and the U.

Now let J be an abelian group in $\underline{A}$, and $u: T(1) \to J$ be a (non-homogeneous) cocycle in $Z^1(U,J)$. Beck [9, pp. 50-51] constructs an object J_u of $\underline{A}^J$ satisfying the conditions below

(5.5) $J_u \approx J$ in $\underline{B}^J$

(5.6) $J_u \approx J_v$ in $\underline{A}^J$ if and only if $cl(u) = cl(v)$ in $H^1(U,J)$

(5.7) If X is in $|\underline{A}^J|$, then $X \approx J_u$ for some u in $Z^1(U,J)$ if and only if X is a principal J-object.

As an object of $\underline{A}$, J_u is the pair (J,ξ_u), with the $\underline{B}$-map $\xi_u: T(J) \to J$ defined to be the composite

$$T(J) \longrightarrow T(1) \times T(J) \xrightarrow{u \times \xi_J} J \times J \xrightarrow{\nu_J} J$$

Here ν_J is the multiplication map of J, and the unlabeled arrow denotes the map with projections $T(t): T(J) \to T(1)$, $1_{T(J)}: T(J) \to T(J)$, with $t: J \to 1$. In order to give to J_u

the structure of a J-object in $\underline{A}$, we need an associative and unitary $\underline{A}$-map $\alpha_u : J_u \times J \to J_u$. To this end, we need a $\underline{B}$-map $J \times J \to J$ on the underlying objects which preserves the triple structure maps; this $\underline{B}$-map is simply ν_J. The necessary calculations, as well as the verification of (5.5) - (5.7), are carried through (in a somewhat more general context) in [9, pp. 50-58].

Now let J be in $|Ab(\underline{A})|$, and u be in $Z^1(U,J)$. Our hypotheses on U and the $\underline{B}$-isomorphism $J_u \approx J$ then guarantee that J_u is f.c. in $\underline{A}$. Thus, by Proposition 5.3, J_u is a coflat Galois J-object in $\underline{A}$. It then follows from (5.6) that the formula

$$\kappa_J(cl(u)) = cl(J_u) \qquad\qquad (u \text{ in } Z^1(U,J))$$

gives a well-defined and one-to-one map of sets $\kappa_J : H^1(U,J) \to \Sigma(J)$.

Now let $\varphi : J \to J'$ be in $Ab(\underline{A})$, u be in $Z^1(U,J)$, and $u' = Z^1(U,\varphi)(u)$ in $Z^1(U,J')$. One then checks easily that $\xi_{u'} T(\varphi) = \varphi \xi_u$ in $\underline{B}$, and thus the $\underline{B}$-map $\varphi : J \to J'$ gives rise to an $\underline{A}$-map $\varphi_u : J_u \to J'_{u'}$. A routine computation then shows that

$$\alpha_{u'}(\varphi_u \times \varphi) = \varphi_u \alpha_u : J_u \times J \to J'_{u'}$$

in $\underline{A}$, whence, by Proposition 2.23, $J'_{u'} \approx \tilde{\varphi}(J_u)$ in $\underline{A}^{J'}$. This means that the diagram below commutes

$$\begin{array}{ccc}
H^1(U,J) & \xrightarrow{\quad \kappa_J \quad} & \Sigma(J) \\
\Big\downarrow {\scriptstyle H^1(U,\varphi)} & & \Big\downarrow {\scriptstyle \Sigma(\varphi)} \\
H^1(U,J') & \xrightarrow{\quad \kappa_{J'} \quad} & \Sigma(J')
\end{array}$$

That is, κ_J is natural in J. The fact that κ_J is a homomorphism is an easy consequence of naturality and the fact that U preserves products; we omit details. This completes the proof of (a).

Turning now to (b), Let J be an abelian group in $\underline{A}$ such that U(J) is in $|Ab(\underline{B})|$. Then U(J) is f.c. in $\underline{B}$, in which case J is f.c. in $\underline{A}$. Furthermore, that E(J) is a set follows from the fact that, given Y in $\underline{B}$, the collection of $\underline{A}$-isomorphism classes of objects X of $\underline{A}$ with $U(X) \approx Y$ is a set; this, in turn, is an immediate consequence of the tripleability of U. We may then conclude that J is an $|Ab(\underline{A})|$. There remains only to prove exactness of the desired sequence.

Now, (5.5) guarantees immediately that $X(U)\kappa_J = 0$. On the other hand, if X is a Galois J-object with $\mathfrak{X}(U)(cl(X)) = 0$ in $\mathfrak{X}(U(J))$, then $U(X) \approx U(J)$ in $\underline{B}^{U(J)}$, whence X is a principal J-object by Proposition 5.3. Thus, by (5.7), $X \approx J_u$ in $\underline{A}^J$ for some u in $Z^1(U,J)$, in which case $cl(X) = \kappa_J\,(cl(u))$ in $\mathfrak{X}(J)$. Since we have already observed that κ_J is a monomorphism, we may conclude that the sequence is exact. This completes the proof of the theorem.

Theorem 5.8. Let $U: \underline{A} \to \underline{B}$ be tripleable, and G be a group in $\underline{A}$ such that every Galois U(G)-object in $\underline{B}$ is trivial. Assume, moreover, that an object X of $\underline{A}$ is f.c. if and only if $\underline{B}(1,U(X)) \neq \emptyset$. Then

(a) G is in $|Gp(\underline{A})|$.

(b) X in $|\underline{A}^G|$ is a coflat Galois G-object if and only if X is a principal G-object.

Proof. (b) Let X be a coflat Galois G-object. Since X is faithful, it is f.c., whence $\underline{B}(1,U(X)) \neq \emptyset$. It follows from Corollary 0.3 that U(X) is faithful, and thus a Galois U(G)-object by Remark 1.3 (c). Therefore $U(X) \approx U(G)$ in $\underline{B}^{U(G)}$, by hypotheses, and so X is a principal G-object by Proposition 5.3.
Conversely, let X in $|\underline{A}^G|$ be a principal G-object. Then $\underline{B}(1,U(X)) \neq \emptyset$, and thus X is f.c. in $\underline{A}$. That X is a coflat Galois G-object then follows from Proposition 5.3.

(a) G is f.c. in $\underline{A}$, since $\underline{B}(1,U(G)) \neq \emptyset$. (a) then follows easily from (b) and the fact that the collection of $\underline{A}$-isomorphism classes of objects Y of $\underline{A}$ with $U(Y) \approx U(G)$ is a set, this fact being a consequence of the tripleability of U. The proof is hence complete.

We shall apply the preceding theorems in our discussion of the extensions of algebraic systems mentioned in the introduction. We shall be interested in the cases listed below.

(5.9) $U: \underline{A} \to$ Sets is the forgetful functor, where

(a) $\underline{A}$ is the category of groups.

(b) $\underline{A}$ is the category of K-algebras, K a commutative ring.

(c) $\underline{A}$ is the category of commutative K-algebras, K as in (b).

(d) $\underline{A}$ is the category of R-modules, R any ring.

We shall defer until later the proof of the theorem below.

Theorem 5.10. Let $U\colon \underline{A} \to$ Sets be any of the cases set forth in (5.9). If X is in $|\underline{A}|$, then the forgetful functor

$$U_X\colon (\underline{A},X) \to (\text{Sets},U(X))$$

satisfies the conditions of Theorem 5.4. Furthermore, if G is any group in $(\underline{A},X)$, then

(a) G is in $|Gp(\underline{A},X)|$.

(b) The Galois G-objects in $(\underline{A},X)$ are precisely the principal G-objects.

(c) Every Galois G-object is coflat.

Finally, the natural transformation $\kappa\colon H^1(U_X,-) \to \mathfrak{X}$ is an equivalence.

Note that, if J is in $|Ab(\underline{A},X)|$, then the cohomology groups $H^n(U_X,J)$ are simply the cohomology groups $H^n(X,J)_X$ of [7].

Corollary 5.11. Let $U\colon \underline{A} \to$ Sets be as in (5.9), and X be in $|\underline{A}|$. Let C be

an X-module in Case (5.9 a)

an X-bimodule in Case (5.9 b)

an X-module in Case (5.9 c)

an R-module in Case (5.9 d)

Then the split (group, algebra, etc.) extension $J = C \times X$ has the structure of an abelian group in $(\underline{A},X)$, and the Galois J-objects are precisely the

(Case 5.9 a) Group extensions of X with kernel C, in the sense of Cartan-Eilenberg [11, Chapter XIV].

(Case 5.9 b) K-algebra extensions of X with kernel C in the sense of Shukla [40

(Case 5.9 c) Commutative K-algebra extensions of X with kernel C in the sense of Andre [1].

(Case 5.9 d) R-module extensions of X by C in the sense of [11,Chapter XIV].

In each case, the group $\mathfrak{X}(J)$ is isomorphic to the group of isomorphism classes of extensions of X by C constructed in each of the references cited above. For example, i case (5.9 c) we obtain an isomorphism $\mathfrak{X}(J) \approx \text{Ext}^1_R(X,C)$.

Proof. This follows immediately from Theorem 5.10 and the analysis in [9] of ab lian groups and principal objects in $(\underline{A},X)$ in each of the above cases.

Remark 5.12. (a) For K an arbitrary commutative ring in Cases 5.9, (b) and (c) the preceding corollary, the extensions need not split as K-modules.

(b) We do not know whether Theorem 5.10 holds for an arbitrary tripleable functor $U: \underline{A} \to$ Sets. In particular, we do not know whether the theorem holds for the case in which $\underline{A}$ is the category of compact spaces and continuous maps.

We turn now to the proof of Theorem 5.10, for which we shall need several lemmas.

Lemma 5.12. Let X be a set and G be a group. Then every Galois $X \times G$ -object in (Sets,X) is trivial.

Proof. Let Y be a Galois $X \times G$ -object. Then, since Y is faithful, we may apply Corollary 0.4 to obtain a map $s: X \to Y$ in (Sets,X). Since $X = (X, 1_X)$ is a terminal object of (Sets,X), the lemma then follows from Corollary 1.14.

Lemma 5.13. Let $U: \underline{A} \to$ Sets be as in (5.9), and X be in $|\underline{A}|$. Then $Y = (Y, p_Y)$ in $|(\underline{A},X)|$ is faithful if and only if $U(p_Y): U(Y) \to U(X)$ is onto.

Proof. If $U(p_Y)$ is onto, then we obtain from Corollary 0.4 a map $s: U(X) \to U(Y)$ in (Sets, U(X)). Thus, by Lemma 5.1, Y is faithful.

Conversely, suppose that Y is faithful. Now, if $f: A \to B$ is an $\underline{A}$-map then it is clear that, in all of the case of (5.9), Im(U(f)) has a natural structure of an object of $\underline{A}$; i.e., there exists a commutative diagram in $\underline{A}$

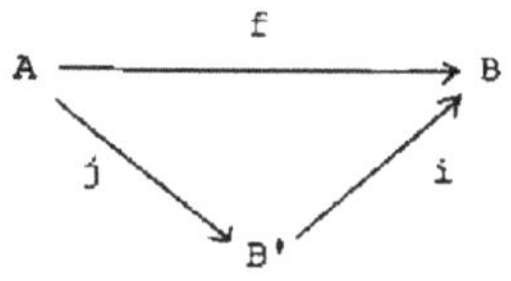

such that U(j) is onto and U(i) is one-to-one. Applying this observation to the $\underline{A}$-map $p_Y: Y \to X$, we may then complete the proof that $U(p_Y)$ is onto by means of an argument quite similar to that of Corollary 0.4; we omit the details.

Lemma 5.14. Let $U: \underline{A} \to$ Sets be as in (5.9), and X be in $|\underline{A}|$. Let Y be in $|(\underline{A},X)|$, with $U(p_Y): U(Y) \to U(X)$ onto. Then Y is coflat in $(\underline{A},X)$.

Proof. Given a c.d.

(5.15) $$A \underset{g}{\overset{f}{\rightrightarrows}} B \xrightarrow{h} C \quad \text{in } (\underline{A},X)$$

we must show that the diagram

(5.16) $$Y \times_X A \overset{Y \times_X f}{\underset{Y \times_X g}{\rightrightarrows}} Y \times_X B \xrightarrow{Y \times_X h} Y \times_X C \quad \text{in } (\underline{A}, X)$$

is likewise a c.d. Applying the forgetful functor $(\underline{A},X) \to \underline{A}$, we obtain the diagrams

(5.17) $$A \overset{f}{\underset{g}{\rightrightarrows}} B \xrightarrow{h} C \quad \text{in } \underline{A}$$

(5.18) $$Y \times_X A \overset{Y \times_X f}{\underset{Y \times_X g}{\rightrightarrows}} Y \times_X B \xrightarrow{Y \times_X h} Y \times_X C \quad \text{in } \underline{A}$$

and we have from Lemma 2.2 that (5.17) is a c.d., and we need only show that (5.18) is a c.d. Now, in all cases $\underline{A}$ possesses coequalizers, and thus we obtain a commutative diagram

(5.19) $$\begin{array}{ccccc} Y \times_X A & \overset{Y \times_X f}{\underset{Y \times_X g}{\rightrightarrows}} & Y \times_X B & \xrightarrow{h'} & C' \\ & & & {\scriptstyle Y \times_X h} \searrow & \downarrow \alpha \\ & & & & Y \times_X C \end{array} \quad \text{in } \underline{A}$$

with the row a c.d. We must then show that α is an isomorphism.

Case 5.9 (a) $\underline{A}$ is the category of groups, with $U\colon \underline{A} \to$ Sets the forgetful functor. It is easily checked that we may assume $C = B/B'$, where B' is the smallest normal subgroup of B generated by all elements of the form $f(a)g(a)^{-1}$ with a in A, and h is the canonical map. Another way of saying this is as follows: Define a (symmetric) relation $\sim$ on B by the condition

(5.20) $b_1 \sim b_2$ if and only if one of the conditions below holds

$$b_2 v g(a) v^{-1} = b_1 v f(a) v^{-1}$$

$$b_2 v f(a) v^{-1} = b_1 v g(a) v^{-1}$$

for some a in A and v in B.

Let $\underset{fg}{\sim}$ be the equivalence relation on B generated by $\sim$. Then $C = B/\underset{fg}{\sim}$, with h the canonical map. Of course, we then have also that $C' = Y \times_X B/\underset{FG}{\sim}$, with $F = Y \times_X f$, $G = Y \times_X g$.

We pause for a remark regarding these constructions.

(5.21) If $b_1 \sim b_2$ in B and (y,b_1) is in $Y \times_X B$, then (y,b_2) is also in $Y \times_X B$.

For, if $b_2vg(a)v^{-1} = b_1vf(a)v^{-1}$, for example, with a in A and v in B, then the equation $p_Bf = p_A = p_Bg$ gives easily that $p_B(b_2) = p_B(b_1) = p_Y(y)$, which proves the assertion.

Now, since h is onto (we suppress the U) and $p_Ch = p_B$, it is trivially verified that $Y \times_X h$ is onto. Thus, in order to complete the proof of this case, we need only show that α is one-to-one; this reduces easily to the following assertion

(5.22) If (y,b_i) $(i = 1,2)$ are in $Y \times_X B$ and $b_1 \widetilde{fg}\, b_2$ in B, then $(y,b_1) \widetilde{FG}\, (y,b_2)$ in $Y \times_X B$.

By (5.21), we may assume without loss of generality that $b_1 \sim b_2$; i.e., they satisfy (5.20) for some a in A, v in B. Since p_Y is onto, we can find u,z in Y with $p_Y(u) = p_B(v)$, $p_Y(z) = p_A(a)$. Then (z,a) and (u,v) are in $Y \times_X A$ and $Y \times_X B$, respectively, and easy computation gives

$$(y,b_2)(u,v)G(z,a)(u,v)^{-1} = (yuzu^{-1},b_2vg(a)v^{-1})$$

$$(y,b_1)(u,v)F(z,a)(u,v)^{-1} = (yuzu^{-1},b_1vf(a)v^{-1})$$

$$(y,b_2)(u,v)F(z,a)(u,v)^{-1} = (yuzu^{-1},b_2vf(a)v^{-1})$$

$$(y,b_1)(u,v)G(z,a)(u,v)^{-1} = (yuzu^{-1},b_1vg(a)v^{-1})$$

Thus $(y,b_2) \widetilde{FG}\, (y,b_1)$, and (5,22) is proved. This establishes the lemma in Case (5.9 a).

Cases (5.9 b)-(5.9 e) The truth of the lemma in these cases follows from arguments similar to that of Case (5.9 a); these we shall omit.

Proof of Theorem 5.10. The tripleability of U is established in [9]; an easy argument then shows that U_X is likewise tripleable. If Y is in $|(\underline{A},X)|$, then Lemmas 5.13 and 5.14 guarantee that Y is f.c. in $(\underline{A},X)$ if and only if $U(p_Y)$: $U(Y) \to U(X)$ is onto. Now, if W is in (Sets,U(X)) and p_W: $W \to U(X)$ is onto, then it is easy to see that W is coflat in (Sets,U(X)); the argument for this bears some resemblance to (but is much simpler than) that of Lemma 5.14. It then follows from Corollary 0.4 that W is f.c. in (Sets,U(X)) if and only if p_W is onto. Therefore, by the preceding remarks, Y is f.c. in $(\underline{A},X)$ if and only if $U_X(Y)$ is f.c. in (Sets,U(X)). We conclude that U_X satisfies the conditions of Theorem 5.4.

By Lemma 5.12, every Galois object of (Sets,U(X)) is trivial. And the preceding remarks, together with the Axiom of Choice, show that Y in $|(\underline{A},X)|$ is f.c. if and only if there exists a map s: U(X) → U(Y) in (Sets,U(X)). Since U(X) is a terminal object of (Sets,U(X)), we obtain that Theorem 5.8 holds for any group G in ($\underline{A}$,X). That is, G is in $|Gp(\underline{A},X)|$ and the coflat Galois G-objects in ($\underline{A}$,X) are precisely the principal G-objects. But, by Lemmas 5.13 and 5.14, every faithful object of ($\underline{A}$,X) is coflat, whence every Galois G-object is coflat. This means that the Galois G-objects in ($\underline{A}$,X) are precisely the principal G-objects. Thus conditions (a)-(c) of Theorem 5.10 hold.

Finally, since (as noted above) every Galois object of (Sets,U(X)) is trivial, we see that the map $\mathfrak{X}(U_X): \mathfrak{X}(J) \to \mathfrak{X}(U_X(J))$ of Theorem 5.4 is the zero map for J in $|Ab(\underline{A},X)|$ whence, by that theorem, $\kappa_J: H^1(U_X,J) \xrightarrow{\approx} \mathfrak{X}(J)$ is an isomorphism. This completes the proof of Theorem 5.10.

We turn next to another special case of Theorem 5.4.

<u>Theorem 5.23.</u> Let $\underline{A}$ be a category with finite products and coequalizers, and X in $|\underline{A}|$ be f.c. Then the functor

$$U = X \times (\ ): \underline{A} \to (\underline{A},X)$$

satisfies the conditions of Theorem 5.4. Furthermore, if J is an abelian group in $\underline{A}$, then there exist isomorphisms

$$H^n(U,J) \approx H^n(X \to 1,\underset{\sim}{J})$$

which are natural in J, where $H^n(X \to 1,\underset{\sim}{J})$ denotes the n'th. Cech cohomology group of the $\underline{A}$-map X → 1 with coefficients in the presheaf $\underset{\sim}{J} = \underline{A}(-,J)$ on $\underline{A}$ [2,p.19]. Viewing these isomorphisms as identifications, we then obtain the exact sequence

$$0 \longrightarrow H^1(X \to 1,\underset{\sim}{J}) \xrightarrow{\kappa_J} \mathfrak{X}(J) \xrightarrow{\mathfrak{X}(U)} \mathfrak{X}(X \times J)$$

if X × J is in $|Ab(\underline{A},X)|$.

Proof. The assumptions on X guarantee, via Remark 2.6, that U is tripleable; this is a consequence of one of the "tripleability theorems" which exist in unpublished work of Beck-Lawvere-Linton. A direct proof can be given.[3]

Now let Y be in $|\underline{A}|$. If Y is f.c. in $\underline{A}$, then we have from Propositions 0.5 and 2.7 that U(Y) = X × Y is f.c. in ($\underline{A}$,X). Conversely, suppose that X × Y is f.c. in ($\underline{A}$,X), and consider the diagram

$$(5.24) \qquad A \underset{g}{\overset{f}{\rightrightarrows}} B \xrightarrow{h} C \quad \text{in } \underline{A}$$

with hf = hg. Since X is f.c. in $\underline{A}$, we obtain from Remark 2.6 that (5.24) is a c.d. if and only if

$$X \times A \underset{X \times g}{\overset{X \times f}{\rightrightarrows}} X \times B \xrightarrow{X \times h} X \times C$$

is a c.d. in $\underline{A}$, which, by Lemma 2.2, is true if and only if it is a c.d. in $(\underline{A},X)$. This will in turn be true if and only if

$$(X \times Y) \times_X (X \times A) \rightrightarrows (X \times Y) \times_X (X \times B) \longrightarrow (X \times Y) \times_X (X \times C)$$

is a c.d. in $(\underline{A},X)$, since $X \times Y$ is f.c. in $(\underline{A},X)$. But, by Lemma 2.2 again, this latter diagram will be a c.d. in $(\underline{A},X)$ if and only if its image in $\underline{A}$ under the forgetful functor $(\underline{A},X) \to \underline{A}$ is a c.d. in $\underline{A}$, this image being

$$X \times (Y \times A) \rightrightarrows X \times (Y \times B) \longrightarrow X \times (Y \times C)$$

But, since X is f.c., this will be a c.d. in $\underline{A}$ if and only if

$$(5.25) \qquad Y \times A \underset{Y \times g}{\overset{Y \times f}{\rightrightarrows}} Y \times B \xrightarrow{Y \times h} Y \times C \quad \text{in } \underline{A}$$

is a c.d. We have proved that (5.24) is a c.d. if and only if (5.25) is, whence Y is f.c. in $\underline{A}$, by Remark 2.6. We may then conclude that U satisfies the conditions of Theorem 5.4.

The natural isomorphisms between the Cech and triple cohomology groups are easily obtained from an examination of the complexes used to define each. The remainder of the theorem is then a consequence of Theorem 5.4, and the proof is complete.

G.S. Rinehart [45] has recently developed a general theory of satellite functors on non-additive categories, which includes within it a theory of extensions of algebraic systems. His definition of extension involves the appropriate specialization of coflatness, as well as of the requirement that $\gamma_X: X \times G \to X \times X$ be an isomorphism.

II. Hopf Algebras and Galois Theory

Stephen U. Chase and Moss E. Sweedler

6. Introduction

In these notes we present a generalization of the fundamental theorem of Galois theory for commutative rings, in which, roughly speaking, the finite group of automorph isms in the classical theory is replaced by a Hopf algebra. Our approach to the subject was inspired primarily by the papers of Chase-Harrison-Rosenberg [12] and Sweedler [42, 43]. The motivation for this work was a hope that results of this type should shed some light on inseparable extensions of fields and ramified extensions of rings.

In Section 7 we set forth the basic concepts with which we shall work, after which we state our form of the fundamental theorem of Galois theory. Then we apply our result to obtain the fundamental theorem, proved in [12], of the Galois theory of separable algebras over a commutative ring.

In Section 8 we summarize, mostly without proof, some results of K. Morita [30] on projective modules and their endomorphism rings. These will be heavily used in later sections of the notes. Our exposition is based on that of H. Bass [8].

Sections 9 and 10 set forth some elementary properties of Hopf algebras and Galois objects. The latter are introduced in Definition 7.3 and form a special case of the categorical Galois objects of Chapter I (see Proposition 4.7). The role which they play i our theory is the role played by the normal separable field extensions in the classical theory.

In Section 11 we prove the fundamental theorem mentioned above and stated in Section 7.

We have tried to make these notes self-contained to the extent that the reader nee not first expose himself to the general theory of Chapter I. In fact, the definition of Galois object, in the particular context discussed here, appears slightly different from that of the general theory. In Section 12, however, we show that the Galois objects introduced in Definition 7.3 are indeed precisely the Galois objects in the category of commutative R-algebras, as described in Section 4. This is a consequence of a technical result which implies, in essence, that the flatness assumption of Definition 7.3 can be removed. In this section we also interpret, in some special cases, the functor X of Section 3.

Throughout the following discussion all rings and modules will be unital. All algebras and coalgebras will, unless specifically stated otherwise, be over a fixed commutative ring R. Unadorned $\otimes$ will mean $\otimes_R$.

7. Definitions, Theorems and Consequences

In our discussion of coalgebras and Hopf algebras, we shall use the definitions and terminology of [27, Chapter VI, Section 9], with the proviso that the grading introduced there be ignored (i.e., all R-modules, coalgebras, etc., will be trivially graded We shall denote by Δ and ε, respectively, the diagonal and counit maps of a coalgebra C When discussing several coalgebras simultaneously, we shall sometimes write $\Delta = \Delta_C$, $\varepsilon = \varepsilon_C$, etc., in order to avoid confusion.

For x in a coalgebra C, we write $\sum_{(x)} x_{(1)} \otimes x_{(2)}$ to denote $\Delta(x)$, $\sum_{(x)} x_{(1)} \otimes x_{(2)} \otimes x_{(3)}$ to denote $(\Delta \otimes 1)\Delta(x)$, etc. If $f: C \otimes \ldots \otimes C \to M$ is an R-module homomorphism, we write

$$\sum_{(x)} f(x_{(1)},\ldots,x_{(n)}) = f(\sum_{(x)} x_{(1)} \otimes \cdots \otimes x_{(n)})$$

In this notation the identity relating ε and Δ becomes

$$\sum_{(x)} \varepsilon(x_{(1)})x_{(2)} = x = \sum_{(x)} x_{(1)}\varepsilon(x_{(2)})$$

A coalgebra C is called <u>cocommutative</u> if, for all x in C

$$\sum_{(x)} x_{(2)} \otimes x_{(1)} = \sum_{(x)} x_{(1)} \otimes x_{(2)}$$

If C is a cocommutative coalgebra we can permute the numerical subscripts arbitrarily in any computation involving $\sum_{(x)} x_{(1)} \otimes \cdots \otimes x_{(n)}$.

A <u>Hopf algebra with antipode</u> is a Hopf algebra A equipped with an R-module homomorphism $\lambda: A \to A$ (called an <u>antipode</u> or <u>inverse</u>) such that

$$\mu(1 \otimes \lambda)\Delta = \eta\varepsilon = \mu(\lambda \otimes 1)\Delta: A \to A ,$$

where $\mu: A \otimes A \to A$ and $\eta: R \to A$ are the multiplication and unit maps, respectively, of the algebra A. λ is both an algebra antiendomorphism and a coalgebra antiendomorphism [26, Section 8]. That is, if $\omega: A \otimes A \to A \otimes A$ is the map which interchanges the factors then $\omega\Delta\lambda = (\lambda \otimes \lambda)\Delta$ and $\lambda\mu\omega = \mu(\lambda \otimes \lambda)$. In particular, if A is a commutative algebra (cocommutative coalgebra) then λ is an algebra (coalgebra) homomorphism.

Now we introduce a type of Hopf algebra with which we shall be almost exclusively concerned in the following sections.

Definition 7.1. (a) A finite Hopf algebra is a Hopf algebra A with antipode which is a finitely generated projective R-module.

(b) A homomorphism f: A → B of finite Hopf algebras is a Hopf algebra homomorphism which preserves antipodes; i.e., $\lambda_B f = f\lambda_A$.

(c) An admissible Hopf subalgebra of a finite Hopf algebra B is the image of a homomorphism f: A → B of finite Hopf algebras with the following property: There exists a homomorphism g: B → A of R-modules such that $gf = 1_A$.

If A is a finite Hopf algebra, then A* is likewise, where $(-)^*$ denotes the functor $\mathrm{Hom}_R(-,R)$. The structure maps of A* are as follows:

$$\mu_{A*} = \Delta_A{}^*: A^* \otimes A^* \approx (A \otimes A)^* \to A^*$$

$$\Delta_{A*} = \mu_A{}^*: A^* \to (A \otimes A)^* \approx A^* \otimes A^*$$

$$\eta_{A*} = \varepsilon_A{}^*: R \to A^*$$

$$\varepsilon_{A*} = \eta_A{}^*: A^* \to R$$

A* is called the dual of A. Clearly A** ≈ A as finite Hopf algebras. If the homomorphism f: A → B of finite Hopf algebras is onto, then f*: B* → A* is one-to-one, and we shall usually identify B* with its image in A*. Note that then B* is an admissible Hopf subalgebra of A*; moreover, every admissible Hopf subalgebra of A* is of this form. Hence, for brevity's sake, we will sometimes say, "Let B* be an admissible Hopf subalgebra of A*".

Let A be a Hopf algebra. An A-object is a pair (S,α), where S is a commutative algebra and α: S → S ⊗ A is an R-algebra homomorphism such that

$$(\alpha \otimes 1)\alpha = (1 \otimes \Delta)\alpha: S \to S \otimes A \otimes A$$

and

$$(1 \otimes \varepsilon)\alpha = 1_S: S \to S \otimes R = S .$$

For brevity, we shall usually denote the pair (S,α) by the symbol S alone. When the map α needs explicit mention, we shall write $\alpha = \alpha_S$. We shall apply to the map α_S the notational device set forth above, and write, for x in S:

$$\alpha_S(x) = \sum_{(x)} x_{(1)} \otimes x_{(2)}$$

$$(\alpha_S \otimes 1)\alpha_S(x) = (1 \otimes \Delta)\alpha_S(x) = \sum_{(x)} x_{(1)} \otimes x_{(2)} \otimes x_{(3)}$$

and so forth. Of course, one must keep in mind that each $x_{(1)}$ is in S, each $x_{(2)}$ is in A, etc.

If A is a finite Hopf algebra and S is an R-module, then we have the following chain of natural isomorphisms

$$\mathrm{Hom}_R(S, S \otimes A) \approx \mathrm{Hom}_R(S, \mathrm{Hom}_R(A^*, S)) \approx \mathrm{Hom}_R(A^* \otimes S, S)$$

the first arising from the fact that A is a finitely generated projective R-module, the second being the well-known adjointness isomorphism. In particular, if S is an A-object, we may apply these isomorphisms to the element α_S of $\mathrm{Hom}_R(S, S \otimes A)$ to obtain a map β_S: $A^* \otimes S \to S$. We shall write $\beta_S(u \otimes x) = u(x)$. It is easy to see that β_S gives rise to an A^*-module structure on S, and the formula below holds

(7.2) $$u(x) = \sum_{(x)} x_{(1)}\langle u, x_{(2)}\rangle \qquad (x \text{ in } S,\ u \text{ in } A^*)$$

where $\langle\ \rangle$: $A^* \otimes A \to R$ denotes the duality pairing. Note further that A^* <u>measures</u> S to S, via the map β_S, as in [43]. Finally, it is trivially verified that, for the special case in which S = A, the left A^*-module structure on A of (7.2) agrees with that induced by the right A^*-module structure on A^* via duality; i.e.

$$\langle u, v(a)\rangle = \langle uv, a\rangle$$

for all a in A and u,v in A^*.

<u>Definition 7.3.</u> Let A be a Hopf algebra, and S be an A-object. We define the algebra homomorphism γ_S: $S \otimes S \to S \otimes A$ by the formula

$$\gamma_S(x \otimes y) = (x \otimes 1)\alpha_S(y) = \sum_{(y)} xy_{(1)} \otimes y_{(2)}$$

S will be called a <u>Galois A-object</u> if the following conditions hold -

(a) S is a faithfully flat R-module.

(b) γ_S: $S \otimes S \to S \otimes A$ is an isomorphism.

Before stating our main theorem on Galois objects, we must introduce several additional concepts. If A is a Hopf algebra, we shall write $I_A = \mathrm{Ker}(\varepsilon_A)$, the <u>augmentation</u>

ideal of A. Now let A be a finite Hopf algebra, and S be an A-object. If

$$w = s_1 \otimes u_1 + \ldots + s_n \otimes u_n$$

in $S \otimes A^*$ and x is in S, we shall write

$$w(x) = s_1u_1(x) + \ldots + s_nu_n(x)$$

an element of S. It is easy to see that this expression is well-defined.

Definition 7.4. Let A be a finite Hopf algebra, and S be a Galois A-object. Let B^* be an admissible Hopf subalgebra of A^*, and T be a subalgebra of S. We shall write $T \to B^*$ to mean that the following condition holds: Given w in $S \otimes A^*$, $w(T) = 0$ if and only if w is in $S \otimes A^*I_{B^*}$.

Finally, if M is a left A-module, with A a Hopf algebra, we write

(7.5) $$M^A = \{m \text{ in } M | am = \varepsilon(a)m \text{ for all a in } A\}$$

Note that, under the conditions of Definition 1.4, S^{B^*} is a subalgebra of S containing T.

In Section 5 we shall prove the following result, which contains our analogue of the fundamental theorem of Galois theory.

Theorem 7.6. Let A be a finite commutative Hopf algebra, and S be a Galois A-object. Then

(a) S is finitely generated faithful projective R-module, and $S^{A^*} = R$.

(b) If B^* is an admissible Hopf subalgebra of A^* and T is a subalgebra of S which is an R-module direct summand of S, then $T \to B^*$ if and only if $T = S^{B^*}$. If these conditions hold, then the T-algebra S is a Galois $T \otimes B$ -object.

(c) If $T_i \to B_i^*$ $(i = 1,2)$ with T_i, B_i^* as in (b), then $T_1 \subseteq T_2$ if and only if $B_2^* \subseteq B_1^*$. In particular, $T_1 = T_2$ if and only if $B_1^* = B_2^*$.

(d) If B_i^* is an admissible Hopf subalgebra of A^*, then $B_1^* \subseteq B_2^*$ if and only if $S^{B_2^*} \subseteq S^{B_1^*}$. In particular, $B_1^* = B_2^*$ if and only if $S^{B_1^*} = S^{B_2^*}$.

Of course, in view of (b), (d) is simply a restatement of (c). A large part of the theorem may be succinctly stated as follows: The correspondence $B^* \rightsquigarrow S^{B^*}$ is a one-to-one lattice-inverting correspondence between the admissible Hopf subalgebras of A^* and certain subalgebras of S. Unfortunately, we have in general no good characterization of the subalgebras of S which arise in this correspondence. However, our result provides enough information to yield, as an easy corollary, the fundamental theorem of Galois theory for separable algebras as given in [12]. We devote the remainder of this section to a discussion of this special case.[4]

Following Auslander-Goldman [5], we call a commutative algebra S separable if S is a projective $S \otimes S$-module. For the case in which R and S are fields, this condition is equivalent to the condition that S be a finite separable extension of R in the usual sense [34, Theorem 1; 11, IX, 7.10].

Definition 7.7. [12, p. 16]. Let $f,g: T \to S$ be homomorphisms of commutative algebras. f and g are called strongly distinct if, for every non-zero idempotent e of S, there exists x in T with $f(x)e \neq g(x)e$.

We refer the reader to [12, p. 17] for the short and essentially self-contained proof of the following lemma.

Lemma 7.8. Let T be a commutative separable algebra, and $f: T \to R$ be an algebra homomorphism. Then there exists a unique idempotent e in T such that $f(e) = 1$ and $xe = f(x)e$ for all x in T. Furthermore, if $f_1,\dots,f_n$ are pairwise strongly distinct algebra homomorphisms from T to R, then the corresponding idempotents $e_1,\dots,e_n$ are pairwise orthogonal and $f_i(e_j) = \delta_{ij}$, the latter denoting the Kronecker delta.

As an immediate consequence of the preceding lemma, we obtain that strongly distinct homomorphisms of a separable algebra are, in some sense, linearly independent.

Corollary 7.9. Let $g_1,\dots,g_n: T \to S$ be algebra homomorphisms, with T separable. Then the following statements are equivalent

(a) $g_1,\dots,g_n$ are pairwise strongly distinct.

(b) If $c_1,\dots,c_n$ are in S and $c_1g_1(x) + \dots + c_ng_n(x) = 0$ for all x in T, then $c_1 = c_2 = \dots = c_n = 0$.

Proof. If $g_1,\dots,g_n$ are pairwise strongly distinct, define $f_i: S \otimes T \to S$ for $i \leq n$ by $f_i(s \otimes x) = sg_i(x)$; then it is easy to see that $f_1,\dots,f_n$ are pairwise strongly distinct S-algebra homomorphisms. Since $S \otimes T$ is a separable S-algebra, we may apply Lemma 7.8 to obtain pairwise orthogonal idempotents $e_1,\dots,e_n$ of $S \otimes T$ with $f_i(e_j) = \delta_{ij}$. Then, for each $i \leq n$, we have that $0 = c_1f_1(e_i) + \dots + c_nf_n(e_i) = c_i$, showing that (a) implies (b). The argument that (b) implies (a) is trivial, and will be omitted.

Now let G be a finite group. We shall denote by RG the group R-algebra of G, a

Hopf algebra with the usual augmentation (i.e., counit) and diagonal. The antipode of RG is defined by $\lambda(\sigma) = \sigma^{-1}$ for σ in G. Note that RG is a finite Hopf algebra; furthermore if H is a subgroup of G, then RH is an admissible Hopf subalgebra of RG. We shall write GR = RG*; note that GR is simply the set of functions from G to R with the pointwise algebra operations, and coalgebra operations and antipode arising in the obvious way from the group operations in G. If S is a GR-object, then it is easy to see that the action of RG = GR* on S arising from (7.2) gives rise to an action of G on S via algebra automorphisms. Conversely, if G acts on an algebra S in this way, then S is a GR-object with $\alpha_S\colon S \to S \otimes GR \approx GS$ defined by the formula

$$\alpha_S(x)(\sigma) = \sigma(x) \qquad\qquad (x \text{ in } S;\ \sigma \text{ in } G)$$

in which case the new G-action on S arising from (7.2) agrees with the original one. Finally, S is a Galois GR-object if and only if it is a Galois extension of R with Galois group G in the sense of Auslander-Goldman [5] and Chase-Harrison-Rosenberg [12]. An easy direct proof of this fact can be obtained by patching together the relevant portions of [12 , Theorem 1.3 e] and the proof of [15 , Lemma 2.5]. Before interpreting Theorem 7.6 in this case, we have the following lemma.

<u>Lemma 7.10.</u> Let G be a finite group, and S be a GR-object. Let T be a subalgebra of S, and $H = \{\sigma \text{ in } G \mid \sigma(x) = x \text{ for all } x \text{ in } T\}$. Then the following statements are equivalent -

(a) $T \to RH$.

(b) The restrictions to T of any two elements of G are either equal or strongly distinct homomorphisms from T to S.

Proof. This is an easy consequence of Corollary 7.9 via a routine computation with coset representatives, which we omit.

Following [12 , Definition 2.1], we shall call a subalgebra T of S <u>G-strong</u> if it satisfies condition (b) of Lemma 1.10.

<u>Theorem 7.11.</u> [12 , Theorem 2.3]. Let G be a finite group, and S be a Galois GR-object. Then S is separable, and there is a one-to-one lattice-inverting correspondence between subgroups of G and separable subalgebras of S which are G-strong. If H is a subgroup of G, then the corresponding subalgebra is $S^H = \{x \text{ in } S \mid \sigma(x) = x \text{ for all } \sigma \text{ in } H\}$.

If T is a separable G-strong subalgebra of S, then the corresponding subgroup of G is $H_T = \{\sigma \text{ in } G | \sigma(x) = x \text{ for all } x \text{ in } T\}$.

Proof. Note first that the diagram below is commutative

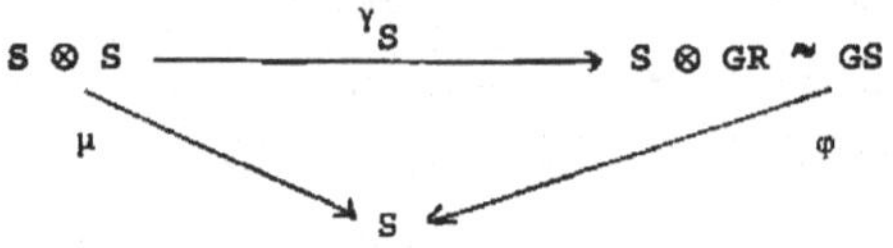

where γ_S is as in Definition 7.3, μ is the multiplication map of S, and $\varphi(u) = u(1)$ for u in GS. Now, it is easy to see that S is a projective GS-module, the module structure arising from φ. Since γ_S is an isomorphism, we may conclude that S is a projective $S \otimes S$-module, and is therefore separable. Then, since S is a finitely generated projective R-module by Theorem 7.6, we may apply [3, Proposition 4.8] or an easy direct argument to obtain that a subalgebra T of S is separable if and only if S is a finitely generated projective T-module, in which case T is a T-module direct summand of S.

Now let H be a subgroup of G. It is then easy to see that $S^{RH} = S^H$. Then $S^H \rightarrow RH$ by Theorem 7.6, and so S^H is G-strong by Lemma 7.10. Furthermore, S is a finitely generated projective S^H-module by parts (a) and (b) of Theorem 7.6, and thus S^H is separable by the remark above.

Conversely, if T is a separable G-strong subalgebra of S then $T \rightarrow RH_T$ by Lemma 7.10. The theorem then follows from Theorem 7.6.

8. The Morita Correspondence

The purpose of this section is to set forth, mostly without proof, some useful properties of projective modules and their endomorphism rings, which will arm us with a powerful technique for our analysis of Galois objects. The original source for this material, in its present generality, is the work of K. Morita [30], although later presentations have been given by Auslander-Goldman [4, Appendix] and, in unpublished work, by H. Bass [8]. Some results of Gabriel [18] also touch on these matters. Our exposition will, for the most part, follow that of [8], except that, for brevity's sake, we will content ourselves with only those parts of the theory necessary for the applications which we have in mind. Thus our discussion will be rather unsystematic, for which we apologize in advance.

Definition 8.1. A Morita Context consists of the following data

(a) R-algebras A and B.

(b) An (A,B)-bimodule P and a (B,A)-bimodule Q, both centralized by R; i.e., $rx = xr$ for all x in P or Q, r in R.

(c) An (A,A)-bimodule homomorphism $\{\ \}: P \otimes_B Q \to A$ and a (B,B)-bimodule homomorphism $[\ ]: Q \otimes_A P \to B$. Given x in P, y in Q, we shall denote the images of $x \otimes y$ and $y \otimes x$, under these mappings, by $\{x,y\}$ and $[y,x]$, respectively. These mappings will be called pairings.

(d) The following equations hold for all x,z in P and y,w in Q

$$\{x,y\}z = x[y,z]$$

$$[y,z]w = y\{z,w\}$$

The Morita context will be called strict if the pairings $\{\ \}$ and $[\ ]$ are surjective.

Note that, under the circumstances introduced above, we obtain an algebra homomorphism $\varphi: B \to \mathrm{End}_A(P)$, and a (B,A)-bimodule homomorphism $\psi: Q \to \mathrm{Hom}_A(P,A)$, defined by the formulae

$$x\varphi(b) = xb \qquad (x \text{ in } P,\ y \text{ in } Q,\ b \text{ in } B)$$

$$x\psi(y) = \{x,y\}$$

(we shall usually write homomorphisms opposite scalars.) In similar fashion we obtain algebra homomorphisms $A \to End_B(P)$, $A \to End_B(Q)$, $B \to End_A(Q)$, and bimodule homomorphisms $P \to Hom_A(Q,A)$, $P \to Hom_B(Q,B)$, $Q \to Hom_B(P,B)$. These homomorphisms will be called the <u>natural</u> mappings arising from the Morita context.

We summarize in the following theorem the most important properties of a strict Morita context. First we introduce some notation. If A is a ring, we shall denote by ${}_A\mathfrak{M}$ the category of left A-modules. If A,B,P,Q, etc., are as in Definition 2.1, and M is a B-submodule of P, we shall write $M^l = \{ y \text{ in } Q \mid [y,M] = 0\}$, a B-submodule of Q which we shall call the left annihilator of M in Q. If N is a B-submodule of Q, we define in similar fashion the right annihilator N^r of N in P. If I is a right ideal of A, we define its left annihilator I^l to be $I^l = \{a \text{ in } A \mid aI = 0\}$, a left ideal of A. If J is a left ideal of A, we define in similar fashion its right annihilator J^r, a right ideal of A.

<u>Theorem 8.2.</u> Let A,B,P,Q,{ },[ ] constitute a strict Morita context. Then

(a) $\{\ \}: P \otimes_B Q \approx A$ and $[\ ]: Q \otimes_A P \approx B$ are isomorphisms.

(b) P and Q are finitely generated projective modules over both A and B.

(c) The natural mappings induce algebra isomorphisms

$$A \approx End_B(P) \approx End_B(Q) \text{ and } B \approx End_A(P) \approx End_A(Q)$$

and bimodule isomorphisms $Q \approx Hom_A(P,A) \approx Hom_B(P,B)$, $P \approx Hom_A(Q,A) \approx Hom_B(Q,B)$.

(d) The functors $Q \otimes_A (-): {}_A\mathfrak{M} \to {}_B\mathfrak{M}$ and $P \otimes_B (-): {}_B\mathfrak{M} \to {}_A\mathfrak{M}$ are isomorphisms of categories, and each is naturally equivalent to the inverse of the other.

(e) The lattice of B-submodules of P(Q) is isomorphic to the lattice of right (left) ideals of A. If M is a B-submodule of P(Q), then the corresponding right (left) ideal of A is $\{M,Q\}(\{P,N\})$. If I is a right (left) ideal of A, then the corresponding B-submodule of P(Q) is IP(QI).

(f) If I is a right (left) ideal of A, then, under the lattice isomorphisms of (e) the B-submodule of Q(P) corresponding to $I^l(I^r)$ is $(IP)^l((QI)^r)$.

Next we consider the special case in which one of the algebras of the Morita context is the ground ring R. First a well-known lemma.

Lemma 8.3. Let I be a finitely generated ideal of R such that $I^2 = I$. Then $I = Re$, with $e^2 = e$.

Proof. Let $x_1,\ldots,x_n$ generate I. Then, since $I^2 = I$, we have that

$$x_i = \sum_{j=1}^{n} x_j a_{ij} \qquad i \leqslant n$$

with a_{ij} in I. Letting α be the $n \times n$ matrix (a_{ij}), we have the vector equation $(x_1,\ldots,x_n)(1-\alpha) = 0$, whence $(x_1,\ldots,x_n)\det(1-\alpha) = 0$. But clearly $\det(1-\alpha) = 1 - e$ with e in I, and so $x_i = x_i e$ for $i \leqslant n$. Since $x_1,\ldots,x_n$ generate I, it follows that $x = xe$ for all x in I, whence $e^2 = e$ and $I = Re$. This completes the proof.

Theorem 8.4. Let the algebras D and R, the left D-module P, the right D-module Q, and the pairings $\{\ \}: P \otimes_R Q \to D$, $[\ ]: Q \otimes_D P \to R$ constitute a Morita context. If P is a faithful R-module and $\{\ \}$ is surjective, then $[\ ]$ is likewise and the Morita context is strict.

Proof. Let $I = [Q,P]$, an ideal of R. Since $\{\ \}$ is surjective, we have $x_1,\ldots,x_n$ in P and $y_1,\ldots,y_n$ in Q such that

$$\sum_{i=1}^{n} \{x_i,y_i\} = 1$$

in D. An easy computation then yields that $x_1,\ldots,x_n$ $(y_1,\ldots,y_n)$ generate P(Q) as R-modules, whence $[y_1,x_1],\ldots,[y_n,x_n]$ generate I. Also, $PI = P$, from which it follows that $I^2 = I$. Thus, by Lemma 2.3, $I = Re$, with $e^2 = e$. Since $PI = P$, we see that $xe = x$, whence $x(1-e) = 0$, for all x in P. But P is a faithful R-module, and so $1 - e = 0$; i.e., $e = 1$ and $I = R$. The theorem follows.

Corollary 8.5. Let P be a finitely generated faithful projective R-module, and $D = \mathrm{End}_R(P)$. View P, in the usual way, as a left D-module. Then R is the center of D, and the evaluation mapping $P \otimes_R \mathrm{Hom}_D(P,D) \to D$ is surjective.

Proof. Let $Q = \mathrm{Hom}_R(P,R)$, and $[\ ]: Q \otimes_D P \to R$ be the duality pairing. Define a pairing $\{\ \}: P \otimes_R Q \to D$ by the formula

$$\{x,y\}z = x[y,z] \qquad (x,z \text{ in } P;\ y \text{ in } Q)$$

It is easily verified that the algebras D and R, the left D-module P, the right D-module Q, and the pairings { }, [ ] constitute a Morita context. Since P is a finitely generated projective R-module, we may apply [11, Ch. VII, Prop. 3.1] to obtain $x_1,\dots x_n$ in P and $y_1,\dots,y_n$ in Q such that, for all x in P

$$x = \sum_{i=1}^{n} x_i[y_i,x] = \sum_{i=1}^{n} \{x_i,y_i\}x$$

whence $\{x_1,y_1\} + \dots + \{x_n,y_n\} = 1$. Thus { } is surjective. Then, by Theorem 8.4, [ ] is likewise surjective and the Morita context is strict. The assertions of the corollary are then easy consequences of Theorem 8.2, part (c).

Theorem 8.6. Let D be an algebra, and S be a commutative subalgebra which is also a left D-module such that $(sv)(x) = sv(x)$ for all s,x in S and v in D. Assume that S is a finitely generated faithful projective R-module, and the natural mapping $D \to \mathrm{End}_R(S)$ is an isomorphism. Let $Q = \{w \text{ in } D / vw = v(1)w \text{ for all } v \text{ in } D\}$, a right ideal in D. Then

(a) The algebras D and R, the left D-module S, the right D-module Q, and the pairings $\{\ \}: S \otimes_R Q \to D$, $[\ ]: Q \otimes_D S \to R$, with

$$\begin{aligned} \{x,w\} &= xw \\ & \qquad\qquad (x \text{ in } S,\ w \text{ in } Q) \\ [w,x] &= w(x) \end{aligned}$$

constitute a strict Morita context.

(b) The unit map $\eta: R \to S$ splits.

Proof. If x,v,w are in S,D,Q, respectively, then $vxw = vx(1)w = v(x)w$ in D. Thus, if s is in S, then $(vw(s))(x) = v(xw(s)) = vxw(s) = v(x)w(s) = (w(s)v)(x)$ for all x in S, whence $vw(s) = w(s)v$ in D. That is, w(s) is in the center of D, which, by Corollary 8.5, is simply R. Thus, $[Q,S] \subseteq R$, from which it follows easily that D,R,S,Q,{ },[ ] constitute a Morita context. Furthermore, a routine computation establishes that the mapping $\varphi: Q \to \mathrm{Hom}_D(S,D)$ defined by the formula

$$x\varphi(w) = xw \qquad (x \text{ in } S,\ w \text{ in } D)$$

is an isomorphism, whence { } is surjective by Corollary 8.5. We may then apply Theorem 8.4 to conclude that the Morita context is strict, establishing (a).

Now, since [ ] is surjective, we may select $s_1,\ldots,s_n$ in S and $w_1,\ldots,w_n$ in Q with $[w_1,s_1] + \ldots + [w_n,s_n] = 1$. Letting $w = w_1s_1 + \ldots + w_ns_n$ in D, one checks easily that the homomorphism $[w,-]: S \to R$ is a left inverse for $\eta: R \to S$. This establishes (b), and completes the proof of the theorem.

9. Elementary Properties of Hopf Algebras and Galois Objects

Throughout this section A will be a finite commutative Hopf algebra.

Proposition 9.1. A is a Galois A-object, with $\alpha_A = \Delta: A \to A \otimes A$.

Proof. A is easily seen to be an A-object. Since A is a projective R-module and $\varepsilon: A \to R$ is onto, it is clear that A is a faithfully flat R-module. Now define the map $\theta: A \otimes A \to A \otimes A$ by the formula

$$\theta(a \otimes b) = \sum_{(b)} a\lambda(b_{(1)}) \otimes b_{(2)} \qquad (a,b \text{ in } A)$$

The properties of the antipode λ, set forth in Section 7, guarantee easily that θ is a two-sided inverse for the map $\gamma_A: A \otimes A \to A \otimes A$. Thus γ_A is an isomorphism, and the proof is complete.

Definition 9.2. [42, Example 7.3] (a) Let S be an A-object. We define an algeb S # A* as follows. As an R-module, S # A* = S ⊗ A*, except that we write s # u rather than s ⊗ u (s in S, u in A*). Multiplication in S # A* is defined by the formula

$$(x \# u)(y \# v) = \sum_{(u)} xu_{(1)}(y) \# u_{(2)}v \qquad (x,y \text{ in } S;\ v,u \text{ in } A^*)$$

with $u_{(1)}(y)$ defined as in (7.2). Routine computations which we omit establish that S # A* is an algebra with unit 1 # 1 (which we shall write simply as 1.) S # A* is called the smash or crossed product of S and A*. Note that S and A* become subalgebras of S # A* via the embeddings $x \to x \# 1$, $u \to 1 \# u$ (x in S, u in A*).

(b) We define a left S # A* -module structure on S by the formula

$$(s \# u)(x) = su(x) \qquad (s,x \text{ in } S;\ u \text{ in } A^*)$$

That the necessary axioms hold is a consequence of (7.2) and the properties of α_S; aga we omit the routine verification.

Theorem 9.3. The statements below are equivalent for any A-object S.

(a) S is a Galois A-object.

(b) S is a finitely generated faithful projective R-module, and the mapping

$\varphi : S \# A^* \to \mathrm{End}_R(S)$, arising from the left $S \# A^*$ -module structure on S, is an isomorphism of algebras.

If these conditions hold, then the unit map $\eta: R \to S$ splits.

Proof. Of course, $\varphi(s \# u)(x) = (s \# u)(x) = su(x)$ for s,x in S and u in A^*.

Assume that S is a Galois A-object; then, by Definition 7.3, the S-algebra map $\gamma_S: S \otimes S \to S \otimes A$ is an isomorphism. Since A is a finitely generated projective R-module, $S \otimes S \approx S \otimes A$ is a finitely generated projective S-module. But since S is a faithfully flat R-module by Definition 7.3, we may then apply [10, Ch. I, Prop. 12, p. 53] to obtain that S is a finitely generated projective R-module.

Next we define the S-module homomorphism $\psi: S \# A^* \to S \otimes S^*$ by the commutative diagram

$$\begin{array}{ccc} S \# A^* & \xrightarrow{\quad\psi\quad} & S \otimes S^* \\ \| & & \| \\ \mathrm{Hom}_S(S \otimes A, S) & \xrightarrow{\mathrm{Hom}_S(\gamma_S, S)} & \mathrm{Hom}_S(S \otimes S, S) \end{array}$$

the vertical S-module isomorphisms arising from the facts that $S \# A^* = S \otimes A^*$ as left S-modules, and A and S are finitely generated projective R-modules. Now let $\{\ \}: S \otimes S^* \to \mathrm{End}_R(S)$ be the usual pairing; i.e. $\{x,v\}y = x\langle v,y\rangle$ for x,y in S and v in S^*, with $\langle\ \rangle: S^* \otimes S \to R$ the duality pairing. $\{\ \}$ is an isomorphism, since S is a finitely generated projective R-module. The formula for γ_S may then be used to show, via an easy computation, that $\{\ \}\psi = \varphi$, whence the latter is an isomorphism. Thus (a) implies (b).

Conversely, assume that (b) holds. Then, since $\{\ \}$ and φ are isomorphisms, ψ is likewise. Therefore $\mathrm{Hom}_S(\gamma_S, S)$ is an isomorphism, by the diagram above. Since S is a finitely generated projective R-module, we may apply the functor $\mathrm{Hom}_S(-,S)$ to obtain that γ_S is an isomorphism. Moreover, by Theorem 8.6, the unit map $\eta: R \to S$ splits, and therefore S is a faithfully flat R-module. It then follows that S is a Galois A-object, whence (b) implies (a) and the proof is complete.

Next we introduce a useful Morita context arising from certain A-objects.

<u>Definition and Remarks 9.4.</u> Let S be an A-object, and assume that $S^{A^*} = R$ (i.e., the unit map $\eta: R \to S$ is one-to-one and $\mathrm{Im}(\eta) = S^{A^*}$). Let $D = S \# A^*$, and

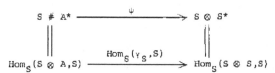

$Q = D^{A^*} = \{w \text{ in } D \mid (1 \# u)w = \varepsilon(u)w = u(1)w \text{ for all } u \text{ in } A^*\}$, a right ideal in D. Define pairings $\{\ \}: S \otimes_R Q \to D$, $[\ ]: Q \otimes_D S \to S^{A^*} = R$ by the formulae

$$\{x,w\} = (x \# 1)u \qquad (x \text{ in } S;\ w \text{ in } Q)$$

$$[w,x] = w(x)$$

(Note that the definition of Q guarantees that [ ] is well-defined.) Then the algebras D and R, the (D,R)-bimodule S, the (R,D)-bimodule Q, and the pairings { },[ ] constitute a Morita Context.

Before analyzing this situation further, we pause for a crucial lemma on Hopf algebras, due to Larson and Sweedler [26].

Lemma 9.5. There exist invertible R-modules I,I' such that $A \approx A^* \otimes I$ ($A \approx I' \otimes A^*$) as left (right) A^*-modules. In particular, A is a projective (left and right) A^*-module.

Proof. We may dualize the isomorphism of the first paragraph following the proof of Proposition 1 of [26], with A playing the role of H, to obtain that $A \approx A^* \otimes I$ as left A^*-modules for some R-module I.[5] (Note that the assumption in [26] that R is a principal ideal domain is unnecessary.) Since A is a finitely generated projective R-module and $A^* = R \oplus \mathrm{Ker}(\varepsilon_{A^*})$, I is likewise a finitely generated projective R-module. A counting of ranks then yields that I has rank one, and is thus invertible [10, Ch. II, §5, Th. 3]. By symmetry, the statement regarding right modules also holds.

Theorem 9.6. The following statements are equivalent for any A-object S.

(a) S is a Galois A-object.

(b) $S^{A^*} = R$, and the Morita context of (9.4) is strict.

If these conditions hold, then S is a projective left module over both $D = S \# A^*$ and A^*, and $Q = JD$ with $J = (A^*)^{A^*}$.

Proof. (a) → (b): Assume that S is a Galois A-object; then by Theorem 9.3, S is a finitely generated faithful projective R-module and the natural map $D \to \mathrm{End}_R(S)$ is an isomorphism. Observe now that $Q = \{w \text{ in } D \mid vw = v(1)w \text{ for all } v \text{ in } D\}$. (b) then follows immediately from Theorem 8.6.

(b) → (a): If (b) holds, then Theorem 8.2 (d) guarantees easily that S is a faithfully flat R-module. That S is a Galois A-object is then an immediate consequence of

Theorem 8.2 (b) and Theorem 9.3.

Assume now that (a), hence (b), hold. Then $\gamma_S: S \otimes S \to S \otimes A$ is an isomorphism. Note that γ_S is a map of right A-comodules; i.e., the diagram below commutes

$$\begin{array}{ccc} S \otimes S & \xrightarrow{\quad \gamma_S \quad} & S \otimes A \\ \downarrow{\scriptstyle S \otimes \alpha_S} & & \downarrow{\scriptstyle S \otimes \Delta_A} \\ S \otimes S \otimes A & \xrightarrow{\quad \gamma_S \otimes A \quad} & S \otimes A \otimes A \end{array}$$

Thus γ_S is an isomorphism of left A*-modules, A* acting on the rightmost factors. Since S is a projective R-module we may then apply Lemma 9.5 to obtain that $S \otimes S$ is a projective left A*-module. Also, the last statement of Theorem 9.3 guarantees that S is a direct summand of $S \otimes S$ as left A*-modules, and is therefore A*-projective. That S is left D-projective is an immediate consequence of (b) and Theorem 8.2 (b).

(b) and Theorem 8.2 also ensure that Q is a projective R-module and $D \approx S \otimes Q$ as left D-modules, hence as left A*-modules. The preceding paragraph then yields that D is a projective left A*-module. That Q = JD then follows from a routine direct sum argument, and the proof of the theorem is complete.

The corollary below is an immediate consequence of Theorems 9.6 and 8.2.

Corollary 9.7. Let S be a Galois A-object, $D = S \# A^*$, and $J = (A^*)^{A^*}$, a right ideal of A*. Then the functors ${}_D\mathcal{M} \to {}_R\mathcal{M}$ and ${}_D\mathcal{M} \to {}_R\mathcal{M}$ defined by

$$M \longrightarrow JM \qquad (M \text{ in } |{}_D\mathcal{M}|)$$

$$N \longrightarrow S \otimes N \qquad (N \text{ in } |{}_R\mathcal{M}|)$$

are isomorphisms of categories and each is naturally equivalent to the inverse of the other. In particular, $M \approx S \otimes JM$ as left D-modules for any M in $|{}_D\mathcal{M}|$.

We end this section with some miscellaneous results which will be useful later.

Proposition 9.8. Let S be a Galois A-object, and $\overline{R}$ be a commutative R-algebra. Then $\overline{R} \otimes S$ is a Galois $\overline{R} \otimes A$ -object.

Proof. It is easy to see that the conditions of Definition 9.1 are preserved by the functor $\overline{R} \otimes (-)$.

Theorem 9.9. Let $f: A^* \to M$ be a homomorphism of left A*-modules, and assume that there exists an R-module homomorphism $g: M \to A^*$ with $gf = 1_{A^*}$. Then there exists an A^*-module homomorphism $h: M \to A^*$ with $hf = 1_{A^*}$.

Proof. We have the exact sequence

$$0 \longrightarrow A^* \xrightarrow{\ f\ } M \xrightarrow{\ p\ } C \longrightarrow 0$$

of A^*-modules, with $C = \mathrm{Coker}(f)$. The sequence splits as a sequence of R-modules. By Lemma 9.5, $A \approx I \otimes A^*$ as right A^*-modules, with I a projective R-module. The usual adjointness relations then yield the following natural equivalences of functors on ${}_{A^*}\mathfrak{M}$:

$$\mathrm{Hom}_{A^*}(-,A^*) \approx \mathrm{Hom}_R(A \otimes_{A^*} (-),R) \approx \mathrm{Hom}_R(I \otimes (-),R)$$

The latter functor manifestly converts the above sequence into an exact sequence of R-modules. Thus we obtain the exact sequence

$$0 \longrightarrow \mathrm{Hom}_{A^*}(C,A^*) \xrightarrow{\mathrm{Hom}_{A^*}(p,A^*)} \mathrm{Hom}_{A^*}(M,A^*) \xrightarrow{\mathrm{Hom}_{A^*}(f,A^*)} \mathrm{Hom}_{A^*}(A^*,A^*) \longrightarrow 0$$

which yields easily the desired conclusion.

10. Subalgebras of Galois Objects

Throughout this section A will be a finite commutative Hopf algebra, and S will be a Galois A-object. As in previous sections, we shall write $D = S \# A^*$. We shall discuss properties of S^{B^*}, for B^* an admissible Hopf subalgebra of A^*. We shall denote by $f\colon A \to B$ the projection of A onto B, the transpose of which is the inclusion map $f^*\colon B^* \to A^*$.

To provide perspective on this situation, we exhibit a characterization of S^{B^*} which can be expressed using only objects and maps in the category of commutative algebras. The proof is an easy computation, which we omit.

Proposition 10.1. The diagram below is an equalizer diagram in the category of commutative algebras

$$S^{B^*} \longrightarrow S \underset{\alpha'_S}{\overset{i}{\rightrightarrows}} S \otimes B$$

where $i(x) = x \otimes 1$, $\alpha'_S = (S \otimes f)\alpha_S$, and the unlabeled map is the inclusion. That is,

$$S^{B^*} = \{x \text{ in } S \mid i(x) = \alpha'_S(x)\} = \{x \text{ in } S \mid \sum_{(x)} x_{(1)} \otimes f(x_{(2)}) = x \otimes 1\}.$$

To simplify notation we shall, at least for awhile, write $T = S^{B^*}$. We shall be interested in the properties of S as a T-algebra. Note that $T \otimes B$ is a finite commutative Hopf T-algebra, and $\mathrm{Hom}_T(T \otimes B, T) \approx T \otimes B^*$. We shall identify the two latter Hopf T-algebras via this natural isomorphism.

We collect in the following proposition some rather trivial, but nevertheless useful, facts regarding S,B,T etc. The proofs are routine consequences of our previous discussion, and will be omitted.

Proposition 10.2. (a) The map $\alpha'_S = (S \otimes f)\alpha_S\colon S \to S \otimes B \approx S \otimes_T (T \otimes B)$ is a homomorphism of T-algebras, and gives the T-algebra S the structure of a $T \otimes B$ -object.

(b) We shall write $D' = S \#_T (T \otimes B^*)$, which is well-defined in view of (a) and Definition 9.2. Then the mapping $j\colon D' \to D$ defined by $j\{x \# (t \otimes u)\} = xt \# u$ (t in T,x in

S, u in B*) is a monomorphism of R-algebras and S-modules, and Im(j) = S # B*, the subalgebra of S # A* generated by all s # u with s in S and u in B*. Furthermore, j converts the D'-module structure on S to the S # B*-module structure on S; i.e., j(w)(x) = w(x) for x in S, w in D'. We shall identify D' with S # B* via the map j.

Note that Proposition 10.2 and (9.4) provide us with a new Morita context, the components of which are the T-algebras D' and T, the (D',T)-bimodule S, and the (T,D')-bimodule $Q' = (D')^{T \otimes B^*} = (D')^{B^*}$ (Q' is, of course, a right ideal in D'). The relevant pairings are { }: $S \otimes_T Q' \to D'$ and [ ]: $Q' \otimes_{D'} S \to T$, where { } is the multiplication pairing and [ ] arises from the D'-module structure of S. That is

$$\{x,w\} = xw \qquad\qquad (x \text{ in } S,\ w \text{ in } Q')$$

$$[w,x] = w(x)$$

Of course, when applying the material of Section 2 to this situation, we must keep in mind that T is now playing the role of R; i.e., the bimodules S and Q' are centralized by T.

Theorem 10.3. The Morita context (D',T,S,Q',{ },[ ]) introduced above is strict. That is, the pairings { },[ ] are surjective.

Proof. Applying Theorem 9.9, with B* playing the role of A*, we obtain a left B*-module homomorphism ζ: A* → B* whose restriction to B* is the identity map of B*. Define now the map θ: D → D' by the formula

$$\theta(x \# u) = x \# \zeta(u) \qquad (x \text{ in } S,\ u \text{ in } A^*)$$

Since ζ is a left B*-module homomorphism, it is easy verified that θ is a left D'-module homomorphism; in particular, $\theta(Q) \subseteq Q'$. Furthermore, the restriction of θ to D' is the identity map of D'.

Now, we have from Theorem 9.6 that the multiplication pairing $S \otimes_R Q \to D$ is surjective; i.e., $x_1w_1 + \ldots + x_nw_n = 1$ for some x_i in S, w_i in Q $(i \leq n)$. We then apply θ to this equation to obtain

$$1 = \theta(1) = \theta(x_1w_1) + \ldots + \theta(x_nw_n) = x_1\theta(w_1) + \ldots + x_n\theta(w_n)$$

with $\theta(w_i)$ in Q'. Thus the multiplication pairing { }: $S \otimes_T Q' \to D'$ is surjective. We may then apply Theorem 8.4 to conclude that the pairing [ ]: $Q' \otimes_{D'} S \to T$ is likewise surjective, and the proof of the theorem is complete.

Corollary 10.4. Let $f: A \to B$ be a surjection of finite Hopf algebras, S be a Galois A-object, and $T = S^{B^*}$. Then the T-algebra S is a Galois $T \otimes B$ -object, the $T \otimes B$ -object structure on S arising from the map $\alpha_S^! = (S \otimes f)\alpha_S: S \to S \otimes_T (T \otimes B)$ of (10.2 a). In particular, S is a finitely generated projective T-module and T is a T-module direct summand of S. Finally, if m is any maximal ideal of R and M is a maximal ideal of T containing m, then $\text{rank}_M(S) = \text{rank}_m(B)$.

Proof. All statements except the last are immediate consequences of Theorems 10.3 and 9.6. Now, if m and M are as above, then $M \cap R = m$, whence we obtain an isomorphism

$$S_M \otimes_{T_M} S_M \approx S_M \otimes_{T_M} (T_M \otimes_{R_m} B_m)$$

of S_M-modules by localizing at M the isomorphism

$$\gamma': S \otimes_T S \to S \otimes_T (T \otimes B)$$

of Definition 7.3. This gives easily that $\text{rank}_M(S_M) = \text{rank}_M(T_M \otimes_{R_m} B_m) = \text{rank}_m B_m$, and the proof is complete.

Corollary 10.5. Let $f: A \to B$ be a surjection of finite Hopf algebras, as in the preceding discussion. Then A^* is a projective (left and right) B^*-module.

Proof. By Proposition 9.1, A is a Galois A-object. Thus, letting $T = A^{B^*}$, we may apply Corollary 10.4 to obtain that the T-algebra A is a Galois $T \otimes B$ -object. Then, by Theorem 9.6, A is a projective left $T \otimes B^*$ -module. Since, by Corollary 10.4, T is a T-module direct summand of A, we also have that T is a projective R-module, and thus $T \otimes B^*$ is a projective left B^*-module. It then follows that A is a projective left B^*-module, whence A^* is a projective left B^*-module by Lemma 9.5. By symmetry, A^* is likewise a projective right B^*-module, completing the proof.

Next we exhibit a result on finite Hopf algebras, the consequences of which will be a crucial importance in the proof of our main theorem.

Theorem 10.6. Let $f: A \to B$ be a surjection of finite Hopf algebras, and $T = A^{B^*}$. Let I be the right ideal of A generated by all elements of A of the form $x - \varepsilon(x)$, with x in T. Then

(a) $I = \mathrm{Ker}(f)$; in particular, I is a two-sided ideal.[6]

(b) $B^* = \{u \text{ in } A^* \mid \langle u, I\rangle = 0\}$, where $\langle\ \rangle: A^* \otimes A \to R$ is the duality pairing.

Proof. (a) If x is in T, then $x \otimes 1 = \sum_{(x)} x_{(1)} \otimes f(x_{(2)})$ in $A \otimes B$, by Proposition 10.1. Applying $\varepsilon \otimes B: A \otimes B \to R \otimes B = B$ to both sides of this equation, we have

$$\varepsilon(x) = \sum_{(x)} \varepsilon(x_{(1)})f(x_{(2)}) = \sum_{(x)} f(\varepsilon(x_{(1)})x_{(2)}) = f(x)$$

whence $f\{x - \varepsilon(x)\} = f(x) - \varepsilon(x) = 0$. Since f is a homomorphism of algebras, it follows that $f(I) = 0$ and $I \subseteq \mathrm{Ker}(f)$.

Now, we have from Theorem 9.9 that there exists a left B^*-module homomorphism $g: A^* \to B^*$ whose restriction to B^* is the identity map of B^*. Set $h = g^*: B \to A$; then $fh = 1_B$ and h is a right B-comodule homomorphism. That is

(10.7) $\qquad (A \otimes f)\Delta_A h = (h \otimes B)\Delta_B : B \to A \otimes B$

Define now a mapping $\theta: A \to A$ by the formula

$$\theta(a) = \sum_{(a)} hf(a_{(1)})\lambda(a_{(2)}) \qquad (a \text{ in } A)$$

Then a routine computation, using (10.7), shows that $(A \otimes f)\Delta_A(\theta(a)) = \theta(a) \otimes 1$ for all a in A, and thus $\mathrm{Im}(\theta) \subseteq T$, by Proposition 10.1. Observe now that $\varepsilon hf = \varepsilon fhf = \varepsilon f = \varepsilon$, from which we see that $\varepsilon\theta = \varepsilon$. Therefore, if a is in A, we have that

$$\sum_{(a)} \varepsilon(\theta(a_{(1)}))a_{(2)} = \sum_{(a)} \varepsilon(a_{(1)})a_{(2)} = a$$

On the other hand

$$\sum_{(a)} \theta(a_{(1)})a_{(2)} = \sum_{(a)} hf(a_{(1)})\lambda(a_{(2)})a_{(3)} = \sum_{(a)} hf(a_{(1)}\varepsilon(a_{(2)})) = hf(a)$$

Thus, if a is in Ker(f), then

$$a = a - hf(a) = \sum_{(a)} \{\varepsilon(\theta(a_{(1)})) - \theta(a_{(1)})\}a_{(2)}$$

is in I since each $\theta(a_{(1)})$ is in T. It then follows that $I = \mathrm{Ker}(f)$, establishing (a).

(b): (a) gives the exact sequence of R-modules

$$0 \to I \xrightarrow{j} A \xrightarrow{f} B \to 0$$

with j the inclusion map. The sequence splits, since B is R-projective. Dualization then gives the exact sequence

$$0 \to B^* \xrightarrow{f^*} A^* \xrightarrow{j^*} I^* \to 0$$

Thus, if u is in A^*, then u is in B^* if and only if $j^*(u) = 0$, which is clearly true if and only if $\langle u, I \rangle = 0$. This establishes (b), and completes the proof of the theorem.

Note that (10.6 a) implies that a surjection $f\colon A \to B$ of finite commutative Hopf algebras is a coequalizer in the category of commutative algebras.

Corollary 10.8. Let A be a finite commutative Hopf algebra, and S be a Galois A-object. Let B_1^*, B_2^* be admissible Hopf subalgebras of A^*. Then $B_1^* \subseteq B_2^*$ if and only if $S^{B_2^*} \subseteq S^{B_1^*}$. In particular, $B_1^* = B_2^*$ if and only if $S^{B_1^*} = S^{B_2^*}$.

Proof. Let $T_i = S^{B_i^*}$ $(i = 1,2)$. It is clear that, if $B_1^* \subseteq B_2^*$, then $T_2 \subseteq T_1$.

Turning to the converse, assume that $T_2 \subseteq T_1$. Let us consider first the special case in which S = A. That $B_1^* \subseteq B_2^*$ in this case is then an immediate consequence of Theorem 10.6 (b).

Now let S be an arbitrary Galois A-object. Then, since the S-algebra isomorphism $\gamma_S\colon S \otimes S \to S \otimes A$ is also a right A-comodule isomorphism, we obtain that it is a left $S \otimes A^*$-module isomorphism (A^* acting, of course, on the second factors). Since

$$S \otimes T_i = (S \otimes S)^{S \otimes B_i^*}$$

by Proposition 10.9 below, it follows that

$$\gamma_S(S \otimes T_i) = (S \otimes A)^{S \otimes B_i^*}$$

whence

$$(S \otimes A)^{S \otimes B_2^*} \subseteq (S \otimes A)^{S \otimes B_1^*} .$$

We then apply the preceding paragraph to obtain that $S \otimes B_1^* \subseteq S \otimes B_2^*$. Since S is a faithfully flat R-module, we may then conclude that $B_1^* \subseteq B_2^*$, completing the proof.

Proposition 10.9. Let $f\colon A \to B$ be a surjection of finite commutative Hopf algebras, and S be a Galois A-object. If $\bar{R}$ is a commutative algebra, then

$$(\bar{R} \otimes S)^{\bar{R} \otimes B^*} = \bar{R} \otimes S^{B^*} .$$

Proof. Let $T = S^{B^*}$. Then, by Corollary 10.4, the T-algebra S is a Galois $T \otimes B$ object, whence the $\overline{R} \otimes T$ -algebra $\overline{R} \otimes S = (\overline{R} \otimes T) \otimes_T S$ is a Galois $\overline{R} \otimes T \otimes B$ -object by Proposition 9.8. Since

$$\overline{R} \otimes T \otimes B^* \approx \mathrm{Hom}_{\overline{R} \otimes T}(\overline{R} \otimes T \otimes B, \overline{R} \otimes T) ,$$

we may apply Theorem 9.6 (b) to obtain that

$$(\overline{R} \otimes S)^{\overline{R} \otimes B^*} = (\overline{R} \otimes S)^{\overline{R} \otimes T \otimes B^*} = \overline{R} \otimes T ,$$

and the proof is complete.

11. Proof of the Main Theorem

The purpose of this section is to exhibit a proof of Theorem 7.6. In preparation for this task, we must first make some remarks regarding duality and annihilators. In what follows, A will be a finite commutative Hopf algebra. As in previous sections, we shall write $I_{A^*} = \mathrm{Ker}(\varepsilon_{A^*})$, and $J_{A^*} = (A^*)^{A^*}$.

Lemma 11.1. Let P be a finitely generated projective left A*-module, and set $Q = P^*$, a right A*-module in the usual way. Let $[\ ]: Q \otimes_{A^*} P \to R$ be the duality pairing. If P_1 is an R-submodule of P, denote by P_1^l the annihilator of P_1 in Q relative to the pairing [ ], as in Section 8. Define similarly $Q_1^r \subseteq P$ for Q_1 an R-submodule of Q. Then $(J_{A^*}P)^l = QI_{A^*}$, and $(QI_{A^*})^r = J_{A^*}P$.

Proof. Let us first consider the special case in which $P = A^*$, $Q = A$. If a is in A, then $[a, J_{A^*}] = [a, J_{A^*}A^*] = [aJ_{A^*}, A^*]$, and so $[a, J_{A^*}] = 0$ if and only if $aJ_{A^*} = 0$. But, by Lemma 9.5, $A \approx M \otimes A^*$ as right A*-modules, with M an invertible R-module. This, together with the definition of J_{A^*}, gives easily that $aJ_{A^*} = 0$ if and only if a is in AI_{A^*}. It then follows that $(J_{A^*})^l = AI_{A^*}$.[7]

On the other hand, if u is in A*, then $[AI_{A^*}, u] = 0$ if and only if $I_{A^*}u = 0$. But, by definition of J_{A^*}, this is true if and only if u is in J_{A^*}. Thus $(AI_{A^*})^r = J_{A^*}$.

We have shown that the lemma holds for the special case $P = A^*$, $Q = A$. The general case then follows from a routine direct sum argument.

In remainder of this section, S will be a Galois A-object. Then, as Theorem 9.6, we have the strict Morita context with algebras $D = S \# A^*$ and R, bimodules S and $Q = D^{A^*} = J_{A^*}D$ (a right ideal of D), and pairings $\{\ \}: S \otimes_R Q \to D$ and $[\ ]: Q \otimes_D S \to R$. Recall that these pairings arise from the multiplication in D and the D-module structure on S, respectively.

We shall use the notation of Section 8 with respect to annihilators. That is, if I is a left (right) ideal of D, we shall denote by $I^r(I^l)$ its right (left) annihilator in D. If M is an R-submodule of S(Q), we shall denote by $M^r(M^l)$ the right (left) annihilator of M in Q(S) relative to the pairing [ ]: $Q \otimes_D S \to R$, as in Theorem 8.2 (f) and Lemma 11.1.

Lemma 11.2. Let B* be an admissible Hopf subalgebra of A*. Then $(J_{B*}D)^l = DI_{B*}$ and $(DI_{B*})^r = J_{B*}D$.

Proof. By Theorem 8.2 (e), there is a one-to-one lattice-preserving correspondence between the R-submodules of S and the right ideals of D such that $J_{B*}D$ corresponds to $J_{B*}(S)$. Also, there is a one-to-one lattice-preserving correspondence between the R-submodules of Q and the left ideals of D, such that DI_{B*} corresponds to QI_{B*}. In addition, by Theorem 8.2 (f), the left ideal $(J_{B*}D)^l$ corresponds to the submodule $(J_{B*}(S))^l$ of Q, and the right ideal $(DI_{B*})^r$ corresponds to the submodule $(QI_{B*})^r$ of S. Finally, by Theorem 8.2 (c), the pairing [ ] induces an R-module isomorphism $Q \approx S^*$.

Now, since A* is a finitely generated projective left B*-module by Corollary 10.5, and S is a finitely generated projective left A*-module by Theorem 9.6, we obtain that S is a finitely generated projective left B*-module. Hence we may apply Lemma 11.1, with B*, S playing the roles of A*,P, respectively, to conclude that $J_{B*}(S)^l = QI_{B*}$ and $(QI_{B*})^r = J_{B*}(S)$. The lemma then follows from the remarks of the preceding paragraph.

Corollary 11.3. Let B* be as in Lemma 11.2. Then $\{S^{B*},Q\} \subseteq J_{B*}D$.

Proof. An easy computation establishes that $\{S^{B*},Q\} \subseteq (DI_{B*})^r$, after which we apply Lemma 11.2.

We turn now to the proof of Theorem 7.6. We shall use the one-to-one lattice-preserving correspondence between the R-submodules of S and the right ideals of D, as set forth in Theorem 8.2 (e). Recall that, if M is an R-submodule of S, then the corresponding right ideal of D is $\{M,Q\} = MQ \subseteq D$; whereas, if I is a right ideal of D, then the corresponding R-submodule of S is I(S).

Let B* be an admissible Hopf subalgebra of A*. An easy computation then yields that $J_{B*}D(S) \subseteq S^{B*}$. Thus, by Corollary 11.3 and the lattice correspondence just mentioned

$$J_{B*}D = \{J_{B*}D(S),Q\} \subseteq \{S^{B*},Q\} \subseteq J_{B*}D$$

from which we may conclude that $\{S^{B*},Q\} = J_{B*}D$ and, using the correspondence once again

$J_{B*}D(S) = S^{B*}$. Furthermore, if w is in D and $w(S^{B*}) = 0$, then $wJ_{B*}D(S) = 0$, in which case $wJ_{B*}D = 0$. Hence, by Lemma 11.2, w is in DI_{B*}. Conversely, it is clear that $DI_{B*}(S^{B*}) = 0$. Thus, if w is in D, then $w(S^{B*}) = 0$ if and only if w is in DI_{B*}. Observe now that the mapping $S \otimes A^* \to D$ defined by $s \otimes u \to s \# u$ is an isomorphism of right A^*-modules. From this it easily follows that $S^{B*} \to B^*$ in the sense of Definition 7.4. Note finally that S is a finitely generated projective S^{B*}-module and S^{B*} is an R-module direct summand of S, by Corollary 10.4.

Now let T be a subalgebra of S which is an R-module direct summand of S, and assume that $T \to B^*$ for some admissible Hopf subalgebra B^* of A^*. If w is in D, then, since $\{T,Q\}(S) = T$, it is clear that $w\{T,Q\} = 0$ if and only if $w(T) = 0$. But, by Definition 7.4 and the remark above, this is true if and only if w is in DI_{B*}. That is, $\{T,Q\}^{l} = DI_{B*}$. It then follows from Lemma 11.2 that $\{T,Q\}^{lr} = J_{B*}D$. But, by Theorem 8.2 and the analogue of Corollary 9.7 for right modules, $\{T,Q\} \approx T \otimes Q$ is a D-module direct summand of D; i.e., the right ideal $\{T,Q\}$ is generated by an idempotent. Therefore $\{T,Q\} = \{T,Q\}^{lr} = J_{B*}D$. Then, by the lattice correspondence and the preceding remarks, $T = \{T,Q\}(S) = J_{B*}D(S) = S^{B*}$. We summarize our findings in the statement below.

(11.4) Let T be a subalgebra of S, and B^* be an admissible Hopf subalgebra of A^*. Then $T = S^{B*}$ if and only if T is an R-module direct summand of S and $T \to B^*$. If these conditions hold, then $J_{B*}D(S) = T$ and $\{T,Q\} = J_{B*}D$.

(11.4), together with Corollary 10.4, yields part (b) of Theorem 7.6. Parts (c) and (d) of the theorem then follow from (b) and Corollary 10.8. Of course, (7.6 a) is simply a restatement of parts of Theorems 9.3 and 9.6, and thus the proof of Theorem 7.6 is complete.

12. Flatness and the Functor X

Let A be a finite Hopf algebra. In view of Proposition 4.6 (c) it is clear that the Galois A-objects introduced in Definition 7.3 are precisely the so-called flat Galois A-objects in the category $\underline{A}$ of commutative R-algebras, in the sense of Section 4. We shall now show that the flatness assumption is irrelevant: Every Galois A-object a la Section 4 is flat, and thus the Galois A-objects of that section are precisely the Galois A-objects of Definition 7.3.

We shall write $J = (A^*)^{A^*}$, as in Section 11.

<u>Lemma 12.1.</u> Let S be a commutative R-algebra, and set $J_S = (S \otimes A^*)^{S \otimes A^*}$. Then the natural map $i\colon S \otimes J \to J_S$ is an isomorphism.

Proof. We have the commutative diagram

$$\begin{array}{ccc}
(S \otimes A) \otimes_S (S \otimes J) \sim S \otimes (A \otimes J) & \xrightarrow{\;S \otimes \pi_A\;} & S \otimes A^* \\
\downarrow {\scriptstyle (S \otimes A) \otimes_S i} & & \| \\
(S \otimes A) \otimes_S J_S & \xrightarrow{\;\pi_{S \otimes A}\;} & S \otimes A^*
\end{array}$$

with $\pi_A\colon A \otimes J \to A^*$, $\pi_{S \otimes A}\colon (S \otimes A) \otimes_S J_S \to S \otimes A^*$ the natural isomorphisms of the first paragraph following Proposition 1 of [26] (with A and $S \otimes A$, respectively, playing the role of H). It follows that $(S \otimes A) \otimes_S i$ is likewise an isomorphism. Since S is an S-module direct summand of $S \otimes A$, we may conclude that i is an isomorphism.

<u>Lemma 12.2.</u> Let S be a Galois A-object a la Section 4; i.e., $\gamma_S\colon S \otimes S \to S \otimes A$ is an isomorphism and S is a faithful object of the category $\underline{A}$. If $D = S \# A^*$, there exist $x_1, \ldots, x_n$ in S and $w_1, \ldots, w_n$ in JD with $x_1 w_1 + \ldots + x_n w_n = 1$.

Proof. We view $S \otimes S \approx S \otimes A$ as an $S \otimes A$ -object in the category $(S, \underline{A})$ of commutative S-algebras. Let $D_S = (S \otimes S) \#_S (S \otimes A^*)$. Then we have the commutative diagram

$$\begin{array}{ccc}
(S \otimes S) \otimes_S (S \otimes JD) \approx S \otimes (S \otimes JD) & \xrightarrow{\quad S \otimes \{\ \} \quad} & S \otimes D \\
\Big\downarrow {\scriptstyle (S \otimes S) \otimes_S k} & & \Big\downarrow {\scriptstyle j} \\
(S \otimes S) \otimes_S (J_S D_S) & \xrightarrow{\quad \{\ \}_S \quad} & D_S
\end{array}$$

with $\{\ \}, \{\ \}_S$ as in Lemma 9.4, j defined by $j(s \otimes (x \# u)) = (s \otimes x) \#_S (1 \otimes u)$, and k induced by j and the map $i: S \otimes J \to J_S$ of Lemma 12.1. j is manifestly an isomorphism, whence k is likewise by Lemma 12.1. Furthermore, since $S \otimes S \approx S \otimes A$ as $S \otimes A$ -objects, we obtain from Theorem 9.6 (with S playing the role of R, etc.) that $\{\ \}_S$, and hence $S \otimes \{\ \}$, are isomorphisms. Since S is a faithful object of $\underline{A}$, we may then apply Proposition 4.6 (a) to conclude that $\{\ \}$is an isomorphism. The lemma follows immediately.

Now let S and A be as in Lemma 12.2, and set $R' = S^{A^*}$. Then $A' = R' \otimes_R A$ is a finite Hopf R'-algebra, and Proposition 10.1 (with A playing the role of B) yields easily that the map $\alpha_S: S \to S \otimes_R A \approx S \otimes_{R'} A'$ is an R'-algebra homomorphism and gives to S the structure of an A'-object in the category $\underline{A}'$ of commutative R'-algebras. In fact, we have the commutative diagram

$$(12.3) \qquad \begin{array}{ccc}
S \otimes_R S & \xrightarrow{\quad \gamma_S \quad} & S \otimes_R A \\
\Big\downarrow {\scriptstyle \kappa} & & \Big\| \\
S \otimes_{R'} S & \xrightarrow{\quad \gamma'_S \quad} & S \otimes_{R'} A'
\end{array}$$

with κ the natural surjection and γ'_S the counterpart of γ_S in the category $\underline{A}'$. Since γ_S is an isomorphism, it follows easily that κ and γ'_S are likewise. Moreover, the mapping $D' = S \#_{R'} (R' \otimes A^*) \to S \# A^* = D$ defined by $s \# (r' \otimes u) \to sr' \# u$ is clearly an isomorphism, and sends J'D' onto JD, where $J' = (R' \otimes A^*)^{R' \otimes A^*}$. This fact, together with Lemma 12.2, yields that the pairing $\{\ \}: S \otimes_{R'} J'D' \to D'$ of Lemma 9.4 (with R' playing the role of R, etc.) is surjective. We may then invoke Lemma 8.4 to obtain the statements below.

<u>Lemma and Remarks 12.4.</u> Let S and A be as in Lemma 12.2, and let R',A',D',J' be as above. Then the Morita Context (D',R',S',J'D',{ },[ ]) of Definition 9.4 (with R' playing

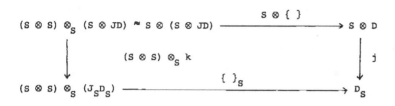

the role of R, etc.) is strict. Thus, by Theorem 9.6, S is a Galois A'-object à la Definition 7.3. Finally, by Theorem 9.3, S is a finitely generated projective R'-module, and the unit map $\nu: R' \to S$ splits.

<u>Theorem 12.5.</u> Let A be a finite Hopf R-algebra, and S be a Galois A-object à la Section 4; i.e., $\gamma_S: S \otimes S \to S \otimes A$ is an isomorphism, and S is a faithful object in $\underline{A}$. Then S is a faithfully flat R-module, and thus is a Galois A-object a la Definition 7.3.

Proof. Let $R' = S^{A^*}$, as above, and

$$R \xrightarrow{\mu} R' \xrightarrow{\nu} S$$

be the canonical maps. We then have the commutative diagram

$$\begin{array}{ccccc} S \otimes_R R & \xrightarrow{S \otimes_R \mu} & S \otimes_R R' & \xrightarrow{S \otimes_R \nu} & S \otimes_R S \\ & \searrow & \downarrow \kappa' & & \downarrow \kappa \\ & & S \otimes_{R'} R' & \xrightarrow{S \otimes_{R'} \nu} & S \otimes_{R'} S \end{array}$$

where κ, κ' are the canonical surjections and the unlabeled arrow represents the natural isomorphism $S \otimes_R R \approx S \approx S \otimes_{R'} R'$. As remarked just after (12.3), κ is an isomorphism. Since, by Lemma 12.4, ν has a one-sided inverse, it follows that $S \otimes_R \nu$ and $S \otimes_{R'} \nu$ are one-to-one, whence κ' is likewise and is thus an isomorphism. Therfore $S \otimes_R \mu$ is an isomorphism. Since S is a faithful object of $\underline{A}$, we may conclude that $\mu: R \to R'$ is an isomorphism. That S is a faithfully flat R-module is then an immediate consequence of Lemma 12.4, and the proof of the theorem is complete.

We denote the remainder of this section to a brief discussion of some examples of the functor $\mathfrak{X}$.

<u>Example 12.6.</u> Let R be a field, π be the Galois group of a separable closure R^s of R, and J be a finite abelian group. Then there exists a natural isomorphism

$$\mathfrak{X}(JR) \approx \mathrm{Hom}_c(\pi, J)$$

where the finite Hopf algebra JR is as in Section 7, and the right-hand side denotes the abelian group of continuous homomorphisms from the compact group π (with the Krull topology) to the discrete group J. The isomorphism may be described explicitly as follows. If $\chi: \pi \to J$ is a continuous homomorphism, let S be the fixed subfield of R^s correspondi to Ker(χ). S is a normal separable extension of R with Galois group naturally isomorphic to $J' = \mathrm{Im}(\chi)$, and is thus a Galois J'R-object in view of the discussion preceding Lemma 7.10. The element of $\mathfrak{X}(JR)$ corresponding to χ is then $\mathrm{cl}(\tilde{\varphi}(S))$, where $\varphi\colon JR \to J'R$ is the homomorphism of Hopf algebras induced by the inclusion $J' \to J$, and $\tilde{\varphi}$ is as in Proposition 4.7 (d).

A similar isomorphism holds for R any connected commutative ring, π then being the fundamental group of R a la Grothendieck (see, e.g., [15, §3]).

Proof. A direct (but rather tedious) verification can be given. A proof of the existence of this isomorphism, from a slightly different point of view, is implicit in [15, §3].

The remarks below can be verified by pursuing further the ideas set forth in Example 4.16 and preceding discussion.

Example 12.7. Let R be a commutative ring, and $Z_n = Z/nZ$. Then there exists an exact sequence

$$U(R) \xrightarrow{n} U(R) \longrightarrow \mathfrak{X}(RZ_n) \longrightarrow \mathrm{Pic}(R) \xrightarrow{n} \mathrm{Pic}(R)$$

with U(R) the multiplication group of invertible elements of R, and Pic(R) the Picard group of R, the elements of which are the isomorphism classes of invertible R-modules. Addition in Pic(R) is induced by $\otimes$. The left-most map in the above sequence raises an element of U(R) to the n'th power; the rightmost map multiplies an element of Pic(R) by n.

Chapter III. Galois Objects and Extensions of Hopf Algebras

by Stephen U. Chase

13. Introduction

Let R be a commutative ring, and A be a commutative cocommutative Hopf R-algebra with antipode which is a finitely generated projective R-module. The principal result of this paper is a natural isomorphism

(13.1) $$X(A) \approx \mathrm{Ext}^1_{\underline{S}}(\underset{\sim}{A}^*, U)$$

where: (a) X(A) is the group of isomorphism classes of Galois A-objects in the category of commutative R-algebras, as in Chapter II,[8]

(b) $\underline{S}$ is the category of abelian sheaves relative to a suitably chosen Grothendieck topology on the category of commutative R-algebras,

(c) $\underset{\sim}{A}^*$ is the sheaf represented by the linear dual A* of A, and

(d) U is the sheaf which assigns to each commutative R-algebra its multiplicative group of invertible elements. The coverings in our topology are essentially "Zariski" coverings i.e., they arise from certain rings of fractions of R.

One of our applications of this result is the following generalization of the well-known Kummer isomorphism of field theory (see, e.g., [37, Chapitre X, §3(b), p. 163].) Let n be prime to the characteristic of a field k, $K = k(\xi)$ with ξ a primitive nth root of 1, U_n be the group of nth roots in 1 in K, and Π and Γ be the Galois groups of $k^s|k$ and $K|k$, respectively with k^s a separable closure of k. Then, for any finite abelian group J of exponent n

(13.2) $$\mathrm{Hom}_c(\Pi, J) \approx \mathrm{Ext}^1_{Z\Gamma}(\mathrm{Hom}_Z(J, U_n), U(K))$$

where the left-hand side denotes continuous homomorphisms from the compact topological group Π to the discrete group J, U(K) is the multiplicative group of invertible elements of K, and the Γ-module structure on $\mathrm{Hom}_Z(J, U_n)$ arises from that of U_n.

The isomorphism (13.1) may be viewed as an analogue, for finite commutative group schemes, of the Weil-Barsotti formula for abelian varieties ([36, Chapitre VII, héorème 6, p. 184] and [31, Chapter III, 18]); it is also related to the Cartier-Shatz ormula of [35, Proposition 1, p. 413]. It reformulates and generalizes some of the work f H. Hasse [23], P. Wolf [44], D.K. Harrison [22], M. Orzech [32,33] and others on Ga-ois algebras and Kummer theory. The proof relies heavily on the coalgebraic techniques f Chapter I, Section 4, and Chapter II, and bears some relation to the method used by hase and Rosenberg in [14].[9]

In Sections 14 and 15 we establish some preliminary results on extensions of pre-heaves and sheaves, the most important of which is the fact that, in our context, an xtension of representable sheaves is likewise representable. Section 16 is devoted to he proof of the main isomorphism (13.1). In Section 17 we consider certain special cases f the theorem. In addition to deriving the isomorphism (13.2), we interpret X(A) - for a commutative, cocommutative Hopf algebra of finite dimension over a field k - in terms f abelian extensions of Hopf k-algebras which "split" as coalgebras. The latter yields ore or less easily, the classical Artin-Schreier theory of extensions of degree p of a ield of characteristic p, in the form provided in [37, Chapitre X, §3(a), p. 163]. hroughout our discussion we shall make uninhibited use of the notation and terminology f Chapter I and II.

It is perhaps appropriate at this point to motivate our approach to this material y a brief consideration of the familiar special case provided by classical Kummer theory. et k be a field of characteristic prime to a given natural number n, and assume that contains all nth roots of 1. We shall scrutinize a normal, separable extension field of k with Galois group J, a finite abelian group of exponent n.

Let V(S) be the set of all x in U(S) such that $x^{-1}\sigma(x)$ is in k for all σ in J. V(S) s clearly a subgroup of U(S) which contains U(k). If x is in V(S), we define a function $_x$ on J by the formula -

$$\varphi_x(\sigma) = x^{-1}\sigma(x) \qquad (\sigma \text{ in } J)$$

$_x$ takes values in U(k); furthermore, we see that, for σ, τ in J -

$$\varphi_x(\sigma\tau) = x^{-1}\sigma\tau(x) = x^{-1}\sigma\{x(x^{-1}\tau(x))\} = (x^{-1}\sigma(x))(x^{-1}\tau(x)) = \varphi_x(\sigma)\varphi_x(\tau)$$

hence φ_x is a homomorphism. Since J has exponent n, it follows that φ_x takes values in the subgroup U_n of nth roots of 1 in k. Also, our assumptions on S guarantee that $\varphi_x(\sigma) = 1$ for all σ in J if and only if x is in U(k).

Suppose now that $\varphi: J \to U_n$ is an arbitrary homomorphism. It is then clear that, with regard to the Galois cohomology of, for example, [37, Chapitre X], φ is a one-cocycle of J with values in U(S), hence by Hilbert's Theorem 90 [37, Proposition 2, p. 15 is a coboundary; i.e., there exists x in U(S) such that $\varphi(\sigma) = x^{-1}\sigma(x)$ for all σ in J. x is then in V(S), and $\varphi = \varphi_x$. We may then conclude that the sequence below is exact -

$$\xi_S: 0 \to U(k) \to V(S) \xrightarrow{\nu_S} \mathrm{Hom}_Z(J,U_n) \to 0$$

(where $\nu_S(x) = \varphi_x$ and the unlabeled arrow denotes the inclusion map), and therefore defines an element of $\mathrm{Ext}_Z^1\,\{\mathrm{Hom}_Z(J,U_n),U(k)\}$.

Recall now that S is a Galois Jk-object, in the language of Section 7, with Jk = (kJ)* the linear dual of the group algebra kJ of J with coefficients in k. Thus it will come as no surprise that the previous considerations, if undertaken with somewhat greater care, produce a homomorphism of abelian groups -

(13.3) $$X(Jk) \to \mathrm{Ext}_Z^1\,\{\mathrm{Hom}_Z(J,U_n),\,U(k)\}$$

which is natural in J.

Let us for the moment take $J = U_n$. Then $\mathrm{Hom}_Z(J,U_n) \approx Z/nZ$, and one obtains immedia ly,via the obvious free resolution of Z/nZ,that the right-hand side of (13.3) is isomorphic to $U(k)/U(k)^n$, with $U(k)^n = \{x^n/x \text{ in } U(k)\}$. A well-known construction then yields an inverse for the map (13.3). That is, if a is in U(k), one sets -

$$S(a) = k[t]/(t^n-a)$$

If α is the image of t in S(a), we define an action of U_n on S(a), via k-algebra automorphisms, by the formula -

$$\zeta(\alpha) = \zeta\alpha \qquad (\zeta \text{ in } U_n)$$

S(a) is then a Jk-object in virtue of Chapter II, Section 7, and is in fact Galois. The inverse map of (13.3) sends cl(a) in $U(k)/U(k)^n$ to cl(S(a)) in X(Jk), where cl( ) denotes the equivalence class of ( ). Of course, S(a) is a field if and only if the equation $t^n - a = 0$ is irreducible in k, in which case S(a)/k is normal and separable with Galois group U_n.

The same construction works for the case in which $J = U_d$, with $d \mid n$. Thus (13.3) is an isomorphism if J is cyclic of exponent n. Naturality and the structure theorem for finite abelian groups then guarantee that (13.3) is an isomorphism for J an arbitrary finite abelian group of exponent n. The relation between (13.3) and (13.1) becomes more apparent if we take note of the natural isomorphism -

$$\mathrm{Hom}_Z(J,U_n) \approx \underline{A}\ (kJ,k) = \underline{A}(Jk^*,k)$$

the middle and right-hand terms denoting k-algebra homomorphisms.

The method of construction of the map (13.1) in the general case is somewhat similar to that outlined above for the map (13.3), except that the desirability of performing base changes renders natural the introduction of sheaves. However, the inverse map is obtained in a rather different way; namely, by a careful scrutiny of the Hopf algebras representing the relevant sheaf extensions.

14. Extensions of Presheaves

In this section we shall be concerned with contravariant functors from a category $\underline{C}$ to the category of abelian groups. We shall adopt the terminology of [2] and will call such a functor an (abelian) presheaf on $\underline{C}$. The category of all presheaves and maps of such (i.e., natural transformations of functors) will be denoted by $\underline{P}(\underline{C})$, and sometimes simply by $\underline{P}$. Thus, in the notation of Chapter I, $\underline{P}(\underline{C}) = (\underline{C}^{op}, Ab)$. It is well-known, and trivially verified, that $\underline{P}(\underline{C})$ is an abelian category, a sequence -

$$F' \to F \to F''$$

being exact if and only if the sequence of abelian groups -

$$F'(X) \to F(X) \to F''(X)$$

is exact for all X in $|\underline{C}|$.

In particular, if F, G are in $|\underline{P}(\underline{C})|$, the abelian group $\mathrm{Ext}^1_{\underline{P}(\underline{C})}(F,G)$ is well-defined, and is functorial in F and G. The elements of $\mathrm{Ext}^1_{\underline{P}(\underline{C})}(F,G)$ are equivalence classe of short exact sequences (s.e.s's) in $\underline{P}(\underline{C})$ of the form -

$$\xi: 0 \to G \to E \to F \to 0$$

such a s.e.s. being called an extension of F by G. The equivalence relation is as follows: If -

$$\xi': 0 \to G \to E' \to F \to 0$$

is another such extension, then $\xi \sim \xi'$ if and only if there exists a presheaf map $E \to E'$ (necessarily an isomorphism) such that the diagram below commutes -

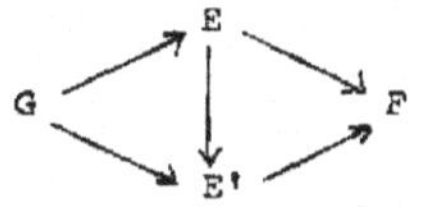

The equivalence class of ξ in $\mathrm{Ext}^1_{\underline{P}(\underline{C})}(F,G)$ will be denoted by $cl(\xi)$. We refer the reader to [20, Chapter XII, §4] for a more detailed discussion of the basic properties of this group.

We shall be particularly interested in $\mathrm{Ext}^1_{\underline{P}(\underline{C})}(F,G)$ for the case in which F and G re representable; say $F = \underline{C}(-,A)$, $G = \underline{C}(-,B)$, with A and B abelian groups in $\underline{C}$. In this ase we shall often write simply $\mathrm{Ext}^1_{\underline{P}(\underline{C})}(A,B)$ for $\mathrm{Ext}^1_{\underline{P}(\underline{C})}(F,G)$, hoping that no confusion ill arise from such abuse of language.

The major portion of the material of this section is well-known (see, e.g. [41, xpose 9].) Hence for several of our results we shall give only brief indications of roof. In the course of this discussion we shall have use for the category $\underline{Q}(\underline{C}) =$ $(\underline{C}^{op},\mathrm{Sets})$ of set-valued presheaves. For brevity, we shall on occasion write $\underline{Q} = \underline{Q}(\underline{C})$, $= \underline{P}(\underline{C})$. Note the existence of the obvious functor from $\underline{P}$ to $\underline{Q}$.

<u>Lemma 14.1.</u> Let $G \overset{\varphi}{\to} E \overset{\varrho}{\to} F$ be a sequence in $\underline{P}(\underline{C})$, and assume that $F = \underline{C}(-,A)$, with n abelian group in $\underline{C}$. Then the following conditions are equivalent -

a)
$$0 \to G \overset{\varphi}{\to} E \overset{\varrho}{\to} F \to 0$$
s a s.e.s. in $\underline{P}(\underline{C})$.

b) The sequence -
$$0 \to G \overset{\varphi}{\to} E \overset{\varrho}{\to} F$$
s exact in $\underline{P}(\underline{C})$, and there exists z in E(A) such that $\varrho(A)(z) = 1_A$ in $\underline{C}(A,A) = F(A)$.

c) There exist $\underline{Q}$-maps $\alpha: F \to E$ and $\beta: E \to G$ such that $\varrho\alpha = 1_F$, $\beta\varphi = 1_G$ and $\varphi\beta + \alpha\varrho = 1_E$.

Proof. That (a) implies (b) is an immediate consequence of a preceding remark escribing exact sequences in $\underline{P}(\underline{C})$. That (c) implies (a) follows from a familiar and outine argument which we omit.

Assume now that (b) holds. The element z in E(A) described above then gives rise a $\underline{Q}$-map $\alpha: F \to E$, defined as follows. If X is in $|\underline{C}|$ and $f: X \to A$ is in $F(X) = \underline{C}(X,A)$, hen $\alpha(X)(f) = E(f)(z)$ in E(X). Since $\varrho(A)(z) = 1_A$ in $F(A) = \underline{C}(A,A)$, it is easy to see hat $\varrho\alpha = 1_F$. Now set $\theta = 1_E - \alpha\varrho : E \to E$, a $\underline{Q}$-map. Then $\varrho\theta = 0$, whence (b) guarantees he existence of a $\underline{Q}$-map $\beta: E \to G$ such that $\varphi\beta = \theta$. (c) follows easily, completing the roof.

Next we show that an extension of representable presheaves is representable.

<u>Theorem 14.2.</u> Let -
$$0 \to G \overset{\varphi}{\to} E \overset{\varrho}{\to} F \to 0$$

be a s.e.s. in $\underline{P}(\underline{C})$. Assume that $\underline{C}$ possesses finite products, and $F = \underline{C}(-,A)$, $G = \underline{C}(-,B)$ with A,B abelian groups in $\underline{C}$. Then $A \times B$ can be given the structure of an abelian group in $\underline{C}$ so that the following conditions hold:

(a) There exists an isomorphism $\zeta\colon E \overset{\approx}{\to} \underline{C}(-,A \times B)$ in $\underline{P}(\underline{C})$.

(b) The diagram below commutes -

$$\begin{array}{ccccc} G & \xrightarrow{\varphi} & E & \xrightarrow{\varrho} & F \\ \| & & \downarrow{\scriptstyle\zeta} & & \| \\ \underline{C}(-,B) & \xrightarrow[\underline{C}(-,\nu)]{} & \underline{C}(-,A \times B) & \xrightarrow[\underline{C}(-,\pi)]{} & \underline{C}(-,A) \end{array}$$

where $\pi\colon A \times B \to A$ is the projection, and $\nu = (\eta_A, 1_B)\colon B = 1 \times B \to A \times B$, with 1 the terminal object of $\underline{C}$ and $\eta_A\colon 1 \to A$ the identity of A.

Proof. Let $\alpha\colon F \to E$ and $\beta\colon E \to G$ be as in Lemma 14.1. Then (14.1c) and a well-known argument show that the $\underline{Q}$-map $(\varrho,\beta)\colon E \to F \times G$ is an isomorphism. The $\underline{Q}$-isomorphism

$$F \times G = \underline{C}(-,A) \times \underline{C}(-,B) \approx \underline{C}(-,A \times B)$$

then guarantee the existence of a $\underline{Q}$-isomorphism $\zeta\colon E \overset{\approx}{\to} \underline{C}(-,A \times B)$. It is then easy to see that $A \times B$ possesses in a unique way the structure of an abelian group in $\underline{C}$ such that ζ is an isomorphism of abelian presheaves. For example, the multiplication map $\mu_{A\times B}\colon (A \times B) \times (A \times B) \to A \times B$ arises, via the Yoneda Lemma, from the composite $\underline{Q}$-map below -

$$\underline{C}(-,A \times B) \times \underline{C}(-,A \times B) \xrightarrow{\zeta^{-1} \times \zeta^{-1}} E \times E \xrightarrow{\nabla} E \xrightarrow{\zeta} \underline{C}(-,A \times B)$$

where ∇ is the addition map; i.e., for x,y in $E(X)$, $\nabla(X)(x,y) = x+y$. (b) then follows from a routine diagram - chasing argument, and the proof of the theorem is complete.

We now turn to an investigation of the behavior of presheaf extensions under a change of category. It will first be necessary to make some remarks on the so-called "factor set" of an extension. Let -

$$\xi\colon 0 \to G \overset{\varphi}{\to} E \overset{\varrho}{\to} F \to 0$$

be a s.e.s. in $\underline{P}(\underline{C})$, and select $\alpha\colon F \to E$, $\beta\colon E \to G$ satisfying the conditions of Lemma 14.1(c). Given X in $|\underline{C}|$, we define a function $f(X)\colon F(X) \times F(X) \to G(X)$ by the formula -

$$f(X)(x,y) = \beta(X)[\alpha(X)(x) - \alpha(X)(x+y) + \alpha(X)(y)]$$

for x,y in F(X). One sees easily from [27, Chapter IV, §4] that f(X) is a factor set corresponding to the extension -

$$\xi(X): 0 \to G(X) \xrightarrow{\varphi(X)} E(X) \xrightarrow{\varrho(X)} F(X) \to 0$$

of abelian groups, and therefore satisfies the identities -

(14.3) $$f(X)(y,z) + f(X)(x,y+z) = f(X)(x,y) + f(X)(x+y,z)$$

(14.4) $$f(X)(x,y) = f(X)(y,x)$$

for x,y,z in F(X), she second holding because the group operation in E(X) is commutative. Now, f(X) is clearly functorial in X, and thus defines a $\underline{Q}(C)$-map $f: F \times F \to G$. (14.3) and (14.4) then assume the form -

(14.5) $$f\nabla_0 + f\nabla_2 = f\nabla_1 + f\nabla_3 \quad \text{in } \underline{Q}(F \times F \times F, G)$$

(14.6) $$f\omega = f \text{ in } \underline{Q}(F \times F, G)$$

where $\omega: F \times F \to F \times F$ interchanges the factors, and $\nabla_i: F \times F \times F \to F \times F$ for $i = 0,\dots,3$ are defined as follows: $\nabla_0(\nabla_3)$ is the projection on the last two (first two) factors, and $\nabla_1 = \nabla \times 1_F$, $\nabla_2 = 1_F \times \nabla$, with $\nabla: F \times F \to F$ the addition map. f will be called a <u>factor set</u> of the extension ξ of F by G.

On the other hand, assume given abelian presheaves F,G on $\underline{C}$, and a $\underline{Q}$-map $f: F{\times}F \to G$ satisfying (14.5) and (14.6). Then, for X in $|\underline{C}|$, the map $f(X): F(X) \times F(X) \to G(X)$ satisfies (14.3) and (14.4), and therefore, by [27, Chapter IV, p. 112], is a factor set of an extension of abelian groups -

$$\xi_f(X): 0 \to G(X) \xrightarrow{\varphi(X)} E(X) \xrightarrow{\varrho(X)} F(X) \to 0$$

where $E(X) = G(X) \times F(X)$ as a set, $\varrho(X)$ is the projection on F(X), $\varphi(X)(u) = (u,0)$ for u in G(X), and addition in E(X) is defined by the formula -

$$(u,x) + (v,y) = (u+v+f(X)(x,y), x + y)$$

for x,y in F(X) and u,v in G(X). This construction is clearly natural in X, and thus we obtain an extension -

$$\xi_f: 0 \to G \xrightarrow{\varphi} E \xrightarrow{\varrho} F \to 0$$

of abelian presheaves on $\underline{C}$. Finally, if f is a factor set of an extension ξ as above, then it is easy to see that $cl(\xi_f) = cl(\xi)$ in $Ext^1_{\underline{P}(\underline{C})}(F,G)$.

Suppose now that $\underline{C}$ possesses finite products, and $F = \underline{C}(-,A)$, with A an abelian group in $\underline{C}$. Since then $F \times F \approx \underline{C}(-,A \times A)$, the Yoneda Lemma tells us that the natural map $G(A \times A) \to \underline{Q}(F \times F,G)$ is an isomorphism, whence a factor set $f: F \times F \to G$ of an extension ξ corresponds to an element $w = w_f$ of $G(A \times A)$. Explicitly, if X is in $|\underline{C}|$ and $x,y: X \to A$ are in $F(X)$, then $f(X)(x,y) = G(z)(w)$ in $G(X)$, where $z = (x,y): X \to A \times A$. In this case we shall, by slight abuse of language, call w a factor set of the extensic ξ. A routine translation of (14.5) and (14.6) then establishes that an element w in $G(A \times A)$ is a factor set of some extension of F by G if and only if the conditions belc hold:

(14.7) $$G(\mu_0)(w) + G(\mu_2)(w) = G(\mu_1)(w) + G(\mu_3)(w) \quad \text{in} \quad G(A \times A \times A) .$$

(14.8) $$G(\tau)(w) = w \quad \text{in} \quad G(A \times A) .$$

Here $\tau: A \times A \to A \times A$ interchanges the factors, $\mu_0: A \times A \times A \to A \times A$ (μ_3) is the projection on the last two (first two) factors, and $\mu_1 = \mu \times 1_A$, $\mu_2 = 1_A \times \mu$, with μ: $A \times A \to A$ the multiplication map of A.

Our sole application of the preceding construction is the proposition below.

<u>Proposition 14.9.</u> Let $\underline{C},\underline{C}'$ be categories with finite products, and $i: \underline{C} \to \underline{C}'$ be a product-preserving functor. Let F',G' be abelian presheaves on $\underline{C}'$, in which case $F = F'i$, $G = G'i$ are abelian presheaves on $\underline{C}$. Define a map -

$$i^*: Ext^1_{\underline{P}(\underline{C}')}(F',G') \to Ext^1_{\underline{P}(\underline{C})}(F,G)$$

as follows: If -

$$\xi': 0 \to G' \xrightarrow{\varphi'} E' \xrightarrow{\varrho'} F' \to 0$$

is a s.e.s. in $\underline{P}(\underline{C}')$, then $i^*(cl(\xi')) = cl(\xi' i)$, with -

$$\xi' i: 0 \to G \xrightarrow{\varphi' i} E' i \xrightarrow{\varrho' i} F \to 0$$

Then i^* is a homomorphism of abelian groups, and is natural in F' and G'. If i is full, faithful, and preserves products, and $F' = \underline{C}'(-,i(A))$ with A an abelian group in $\underline{C}$, then i^* is an isomorphism.

Proof. Routine calculations establish the first assertion. As for the second, observe first that $i(A)$ is an abelian group in $\underline{C}'$, since i preserves products; thus the equation $F' = \underline{C}'(-,i(A))$ makes sense. In this case we have that $F = F'i = \underline{C}'(i(-),i(A)) \approx \underline{C}(-,A)$, since i is full and faithful.

Now let -

$$\xi\colon\ 0 \to G \xrightarrow{\varphi} E \xrightarrow{\varrho} F \to 0$$

be a s.e.s. in $\underline{P}(\underline{C})$, and w in $G(A \times A) = G'i(A \times A)$ be a factor set corresponding to ξ, as in the preceding discussion. Letting w' be the image of w under the natural isomorphism $G'i(A \times A) \approx G'(i(A) \times i(A))$, we have easily from (14.7) and (14.8) that w' is a factor set corresponding to an extension -

$$\xi'\colon\ 0 \to G' \xrightarrow{\varphi'} E' \xrightarrow{\varrho'} F' \to 0$$

in $\underline{P}(\underline{C}')$, and $i^*(\mathrm{cl}(\xi')) = \mathrm{cl}(\xi)$. Therefore i^* is onto.

Finally, let ξ' be an extension of F' by G', as above, and assume that $i^*(\mathrm{cl}(\xi')) = \mathrm{cl}(\xi' i) = 0$. Then, by [27, Proposition 4.2, p. 368], there is a $\underline{P}(\underline{C})$-map $\alpha\colon F \to E'i$ such that $(\varrho' i)\alpha = 1_F$. An argument similar to (but easier than) the preceding then produces a $\underline{P}(\underline{C}')$-map $\alpha'\colon F' \to E'$ such $\varrho'\alpha' = 1_{F'}$, whence, again by [27, Proposition 4.2, p. 368], $\mathrm{cl}(\xi') = 0$ in $\mathrm{Ext}^1_{\underline{P}(\underline{C}')}(F',G')$ and i^* is one-to-one. Thus i^* is an isomorphism and the proof is complete.

<u>Corollary 14.11.</u> Given a diagram of categories and functors -

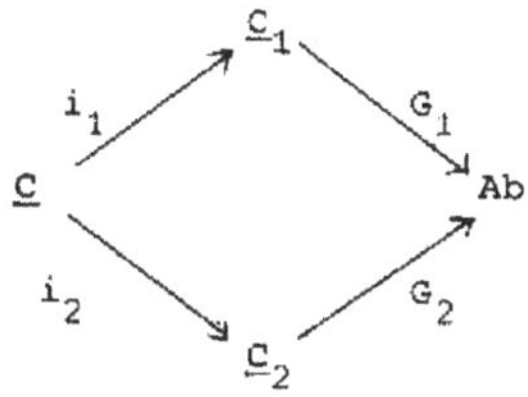

with G_k in $\underline{P}(\underline{C}_k)$ $(k = 1,2.)$ Let $\sigma\colon G_1 i_1 \approx G_2 i_2$ be a $\underline{P}(\underline{C})$-isomorphism, and assume that i_1, i_2 are full, faithful, and preserve products. Let A be an abelian group in $\underline{C}$, and set $F = \underline{C}(-,A)$, $F_k = \underline{C}_k(-,i_k(A))$ in $\underline{P}(\underline{C})$, $\underline{P}(\underline{C}_k)$, respectively. Then there is a unique isomorphism $\kappa\colon \mathrm{Ext}^1_{\underline{P}(\underline{C}_1)}(F_1,G_1) \approx \mathrm{Ext}^1_{\underline{P}(\underline{C}_2)}(F_2,G_2)$ rendering the diagram below commutative -

$$\begin{array}{ccc}
\mathrm{Ext}^1_{\underline{P}(\underline{C}_1)}(F_1,G_1) & \xrightarrow{\kappa} & \mathrm{Ext}^1_{\underline{P}(\underline{C}_2)}(F_2,G_2) \\
\downarrow i_1^* & & \downarrow i_2^* \\
\mathrm{Ext}^1_{\underline{P}(\underline{C})}(F,G_1 i_1) & \xrightarrow[\mathrm{Ext}^1_{\underline{P}(\underline{C})}(F,\sigma)]{} & \mathrm{Ext}^1_{\underline{P}(\underline{C})}(F,G_2 i_2)
\end{array}$$

with i_k^* as in Propostiion 14.9.

We end this section with an example which will be useful in the computations of Section 17.

<u>Example 14.12.</u> Let Γ be a group, and $\underline{C}$ be the category of sets equipped with left Γ-action, and maps of such. Note that an abelian group in $\underline{C}$ is simply a Γ-module; hence Γ-modules A and C give rise to abelian presheaves $F = \underline{C}(-,A)$ and $G = \underline{C}(-,C)$ on $\underline{C}$. We claim that there exists a monomorphism of abelian groups -

$$\theta\colon \mathrm{Ext}^1_{\underline{P}(\underline{C})}(F,G) \to \mathrm{Ext}^1_{Z\Gamma}(A,C)$$

satisfying the following condition: Given -

$$\xi\colon 0 \to G \xrightarrow{\varphi} E \xrightarrow{Q} F \to 0$$

a s.e.s. in $\underline{P}(\underline{C})$, then $\theta(\mathrm{cl}(\xi)) = \mathrm{cl}(\bar{\xi})$, where -

$$\bar{\xi}\colon 0 \to C \xrightarrow{\bar{\varphi}} B \xrightarrow{\bar{Q}} A \to 0$$

is any s.e.s. of Γ-modules such that there exists a $\underline{P}(\underline{C})$-isomorphism $\zeta\colon E \cong \underline{C}(-,B)$ rendering the diagram below commutative -

(14.13)
$$\begin{array}{ccccc}
G & \xrightarrow{\varphi} & E & \xrightarrow{Q} & F \\
\| & & \downarrow \zeta & & \| \\
\underline{C}(-,C) & \xrightarrow[\underline{C}(-,\bar{\varphi})]{} & \underline{C}(-,B) & \xrightarrow[\underline{C}(-,\bar{Q})]{} & \underline{C}(-,A)
\end{array}$$

If $\mathrm{Ext}^1_{Z\Gamma'}(Z,C) = 0$ for every subgroup Γ' of Γ, then θ is an isomorphism.

Proof. Note first that, given ξ as above, Theorem 14.2 guarantees the existence of $\bar{\xi},\zeta$ satisfying the desired conditions. The Yoneda Lemma and a routine argument then establish that θ is a well-defined homomorphism of groups. If $\theta(\mathrm{cl}(\xi)) = 0$, then choose ng $\bar{\xi}$ as above, we obtain a Γ-module homomorphism $\bar{\alpha}$: $A \to B$ such that $\bar{\varrho}\bar{\alpha} = 1_A$. But then, efining the $\underline{P}(\underline{C})$-map α: $F \to E$ to be the composite -

$$F = \underline{C}(-,A) \xrightarrow{\underline{C}(-,\bar{\alpha})} \underline{C}(-,B) \xrightarrow{\xi^{-1}} E$$

e obtain from (14.13) that $\varrho\alpha = 1_F$, whence $cl(\xi) = 0$ in $Ext^1_{\underline{P}(\underline{C})}(F,G)$. Therefore θ is a onomorphism.

Assume now that $Ext^1_{Z\Gamma'}(Z,C) = 0$ for every subgroup Γ' of Γ. Express A as a disjoint nion of Γ-orbits -

$$A = U_i \ Orb(a_i)$$

ith $Orb(a_i) = \{\gamma a_i \text{ in } A/\gamma \text{ in } \Gamma\}$. Let -

$$\Gamma_i = \{\gamma \text{ in } \Gamma/\gamma a_i = a_i\}$$

subgroup of Γ. If -

$$\bar{\xi}: 0 \to C \xrightarrow{\bar{\varphi}} B \xrightarrow{\bar{\varrho}} A \to 0$$

s a s.e.s. of Γ-modules, the exact cohomology sequence -

$$\begin{array}{ccccc} H^0(\Gamma_i,B) & \xrightarrow{H^0(\Gamma_i,\bar{\varrho})} & H^0(\Gamma_i,A) & \longrightarrow & H^1(\Gamma_i,C) \\ \| & & \| & & \| \\ B^{\Gamma_i} & \xrightarrow{\bar{\varrho}} & A^{\Gamma_i} & \longrightarrow & 0 \end{array}$$

with $H^*(\Gamma_i,-) = Ext^*_{Z\Gamma_i}(Z,-)$) guarantees the existence of b_i in B with $\bar{\varrho}(b_i) = a_i$ and $b_i = b_i$ for all γ in Γ_i. We use the Axiom of Choice to select such a b_i for each i, nd then define $\bar{\alpha}$: $A \to B$ by the formula $\bar{\alpha}(\gamma a_i) = \gamma b_i$ for γ in Γ. An easy computation scertains that $\bar{\alpha}$ is a well-defined $\underline{C}$-map (not a homomorphism of Γ-modules) and $\bar{\alpha} = 1_A$.

Now since $\bar{\xi}$ is a s.e.s. of Γ-modules, it is trivially verified that the sequence -

$$0 \to G \xrightarrow{\varphi} E \xrightarrow{\varrho} F$$

s exact in $\underline{P}(\underline{C})$, with $E = \underline{C}(-,B)$, $\varphi = \underline{C}(-,\bar{\varphi})$, $\varrho = \underline{C}(-,\bar{\varrho})$. Moreover, we have just contructed $\bar{\alpha}$ in E(A) such that $\varrho(A)(\bar{\alpha}) = 1_A$ in $\underline{C}(A,A) = F(A)$. We may then apply Lemma 4.1(b) to conclude that -

$$\xi: 0 \to F \xrightarrow{\varphi} E \xrightarrow{\varrho} F \to 0$$

is a s.e.s. in $\underline{P}(\underline{C})$, and clearly $\theta(cl(\xi)) = cl(\overline{\xi})$. We have shown that θ is onto, and thus an isomorphism, completing the proof.

15. Extensions of Sheaves

In the remainder of this paper R will be a commutative ring, and $\underline{A}$ will be the category of commutative R-algebras and homomorphisms of such. Unadorned $\otimes$ will mean $\otimes_R$. We shall denote by $\underline{C}$ the category of cocommutative R-coalgebras, and by $\underline{H}$ the category of commutative, cocommutative Hopf R-algebras with antipode. Recall that H in $|\underline{H}|$ is called a <u>finite</u> Hopf algebra if it is a finitely generated projective R-module; these form a full subcategory $\underline{H}_0$ of $\underline{H}$. If T is in $|\underline{A}|$ we shall write $\underline{A}(T)$, $\underline{C}(T)$, $\underline{H}(T)$, $\underline{H}_0(T)$ for the categories of commutative T-algebras, cocommutative T-coalgebras, etc. Observe that, as remarked in Section 4, $\underline{H}$ may be viewed as the category of either abelian groups in $\underline{C}$ or abelian cogroups in $\underline{A}$.

We now introduce the appropriate setting for our work. We shall call T in $|A|$ a <u>covering</u> of R if the following condition holds -

(15.1) $$T = \prod_{i=1}^{r} R_{x_i}$$

for some $x_1,\ldots,x_r$ in R (depending on T) such that $x_1 + \ldots + x_r = 1$ and no x_i is in the Jacobson radical of R. More generally, given S in $|A|$, we define a covering of S to be an $\underline{A}$-map $S \to S \otimes T$, with T a covering of R in the sense of (15.1) and the map being $s \to s \otimes 1$. Note that $S \otimes T$ is a faithfully flat S-module, by [10, Proposition 3, p. 137 One checks easily that this collection of $\underline{A}$-maps satisfies axioms dual to those of a Grothendieck topology (for the definition and elementary properties of which we refer the reader to [2]), and thus gives rise to a Grothendieck topology on $\underline{A}^{op}$. We may then consider the category $\underline{S}$ of abelian sheaves on this topology. An object of $\underline{S}$ is a (co-variant) functor $F: \underline{A} \to Ab$ such that for any covering $d: S \to S \otimes T$, the sequence below is exact -

(15.2) $$0 \to F(S) \xrightarrow{F(d)} F(S \otimes T) \xrightarrow{F(d^0)-F(d^1)} F(S \otimes T \otimes T)$$

with $d^i\colon S \otimes T \to S \otimes T \otimes T$ $(i = 0,1)$ defined by $d^0(s \otimes t) = s \otimes 1 \otimes t$, $d^1(s \otimes t) = s \otimes t \otimes 1$. A map in $\underline{S}$ is simply a natural transformation of functors; i.e., $\underline{S}$ is a full subcategory of the category $\underline{P} = \underline{P}(\underline{A}^{op}) = (\underline{A}, Ab)$ of abelian presheaves on $\underline{A}^{op}$.

$\underline{S}$ is an abelian category, and a sequence -

$$0 \to F' \xrightarrow{\varphi} F \xrightarrow{\varrho} F'' \to 0$$

in $\underline{S}$ is a short exact sequence if and only if the following conditions hold -

(15.3a) The sequence $0 \to F'(S) \xrightarrow{\varphi(S)} F(S) \xrightarrow{\varrho(S)} F''(S)$

is exact in Ab for any S in $|\underline{A}|$. This means simply that the sequence -

$$0 \to F' \xrightarrow{\varphi} F \xrightarrow{\varrho} F''$$

is exact in $\underline{P}$.

(15.3b) Given any S in $|\underline{A}|$ and x'' in $F''(S)$, there is a covering $d\colon S \to S \otimes T$ and an x in $F(S \otimes T)$ such that $\varrho(S \otimes T)(x) = F''(d)(x'')$.

For the proofs of these facts we refer to [2, Chapter II, Section 1]. Note finally that, since $\underline{S}$ is an abelian category, we may define $Ext^*_{\underline{S}}(-,-)$ as, for example, in [27, Chapter XII]. In particular, if F and G are abelian sheaves, then $Ext^1_{\underline{S}}(F,G)$ may be viewed as the abelian group of equivalence classes of s.e.s.'s in $\underline{S}$, the equivalence relation being that described at the beginning of Section 14.

Although the proposition below is well-known, we sketch a proof for the convenience of the reader.

Proposition 15.4. If H is in $|\underline{H}|$, then $F = \underline{A}(H,-)$ is in $|\underline{S}|$.

Proof. That F is an abelian presheaf is an immediate consequence of the fact that H is an abelian cogroup in $\underline{A}$. Let $d\colon S \to S \otimes T$ be a covering. Since T is a faithfully flat R-module, we have from [13, Lemma 3.8, p.50] that the diagram below is an equalizer diagram in $\underline{A}$ -

$$S \xrightarrow{d} S \otimes T \underset{d^1}{\overset{d^0}{\rightrightarrows}} S \otimes T \otimes T$$

with d^i $(i = 0,1)$ as in (15.2). An easy diagram - chasing argument then shows that F satisfies (15.2), and is hence a sheaf.

<u>Notation 15.5.</u> If H is in $|\underline{H}|$, we shall write $\underset{\sim}{H} = \underline{A}(H,-)$, the sheaf represented by H. If φ: H → H' in $\underline{H}$, we shall denote the induced map of sheaves by $\underset{\sim}{\varphi}: \underset{\sim}{H} \to \underset{\sim}{H}'$.

We turn now to extensions (i.e., s.e.s.'s) of sheaves. The important result below says, in essence, that an extension of representable sheaves is representable.

<u>Proposition 15.6.</u> Let -

$$0 \to \underset{\sim}{C} \to E \to \underset{\sim}{A} \to 0$$

be a s.e.s. in $\underline{S}$, with A,C in $|\underline{H}|$. Then there exists H in $|\underline{H}|$, $\underline{H}$-maps -

$$A \xrightarrow{\varrho} H \xrightarrow{\varphi} C$$

and an isomorphism $\zeta: \underset{\sim}{H} \overset{\approx}{\to} E$ of sheaves such that the diagram below commutes -

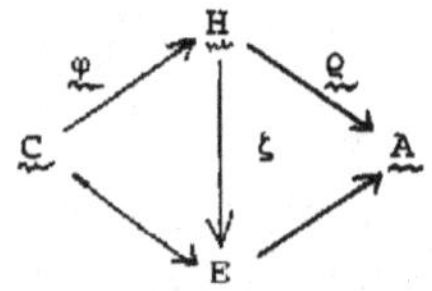

Proof. The proof of the assertion above is an exercise in the theory of descent, and may be accomplished by the techniques of [20] or [13, §4] (see also [31, Proposition 17.4, p. III. 17-6] for a proof of the same result in a somewhat different context.) We shall provide only a brief sketch of the argument, omitting most details.

Given T in $|\underline{A}|$, we have the $\underline{A}$-maps -

$$R \xrightarrow{\varepsilon} T \underset{\varepsilon_1}{\overset{\varepsilon_0}{\rightrightarrows}} T^2$$

with $T^2 = T \otimes T$, ε the unit map, and $\varepsilon^0(t) = 1 \otimes t$, $\varepsilon^1(t) = t \otimes 1$. Setting $\underline{A}' = \underline{A}(T)$, $\underline{A}'' = \underline{A}(T^2)$, we then obtain the following diagrams of categories and functors -

$$\underline{A}'' \underset{\sigma_1}{\overset{\sigma_0}{\rightrightarrows}} \underline{A}' \xrightarrow{\sigma} \underline{A}$$

$$\underline{A} \xrightarrow{\delta} \underline{A}' \underset{\delta_1}{\overset{\delta^0}{\rightrightarrows}} \underline{A}''$$

For example, if T'' is in $|\underline{A}''|$, then $\sigma_i(T'')$ is simply T'' viewed as a T-algebra via

^{i}: $T \to T^2$; whereas, if T' is in $|\underline{A}'|$, then $\delta^i(T') = T' \otimes_T T^2$, T^2 again viewed as a T-lgebra via ε^i. Note that δ and δ^i are left adjoints of σ and σ_i, respectively.

Now, (15.3b) guarantees the existence of a covering $d = 1 \otimes \varepsilon$: $A \to A \otimes T$ of A such hat the image of the map $E(A \otimes T) \to \underset{\sim}{A}(A \otimes T) = \underline{A}(A, A \otimes T)$ contains d. Viewing $E\sigma$, $\underset{\sim}{A}\sigma$, σ as abelian presheaves on the category $(\underline{A}')^{op}$, we note, by the adjointness mentioned bove, that $\underset{\sim}{A}\sigma$ and $\underset{\sim}{C}\sigma$ are represented by $A \otimes T$ and $C \otimes T$, respectively. Thus the image f the map $E\sigma(A \otimes T) \to \underset{\sim}{A}\sigma(A \otimes T) \approx \underline{A}'(A \otimes T, A \otimes T)$ contains the identity map of $A \otimes T$, hence the sequence -

$$0 \to \underset{\sim}{C}\sigma \to E\sigma \to \underset{\sim}{A}\sigma \to 0$$

s an exact sequence of presheaves on $(\underline{A}')^{op}$, by Lemma 14.1(b); moreover, by Theorem 14.2, σ is represented by an object H' of $\underline{H}(T)$.

Since $\varepsilon^0\varepsilon = \varepsilon^1\varepsilon$: $R \to T^2$, we have that $\sigma\sigma_0 = \sigma\sigma_1$: $\underline{A}'' \to \underline{A}$. This equality, together ith adjointness and the definition of H', gives the chain of natural isomorphisms -

$$\underline{A}''(\delta^0(H'),-) \approx E\sigma\sigma_0 = E\sigma\sigma_1 \approx \underline{A}''(\delta^1(H'),-)$$

he Yoneda Lemma then guarantees that the resulting isomorphism is induced by an iso-orphism ω: $\delta^1(H') \overset{\approx}{\to} \delta^0(H')$ of Hopf T^2-algebras. Recalling that $\delta^i(H') = H' \otimes_T T^2$ (the ^{2}-algebra structure being induced by ε^i: $T \to T^2$), we set -

$$H = \{x \text{ in } H' \;/\; \omega[x \otimes (1 \otimes 1)] = x \otimes (1 \otimes 1) \text{ in } \delta^0(H')\}$$

t can then be shown (using, e.g., the techniques of [13, §4]) that H is a Hopf R-algebra nd represents the functor E; i.e., there is an isomorphism ζ: $\underset{\sim}{H} \overset{\approx}{\to} E$ of sheaves. The oneda Lemma then implies the existence of sheaf maps ϱ and φ satisfying the desired onditions.

One of the facts elicited during the course of the preceding discussion will be use-ul later; we restate it below for convenient reference.

Remark 15.7. Let $A \xrightarrow{\varrho} H \xrightarrow{\varphi} C$ be a sequence in $|\underline{H}|$ such that

$$0 \to \underset{\sim}{C} \xrightarrow{\underset{\sim}{\varphi}} \underset{\sim}{H} \xrightarrow{\underset{\sim}{\varrho}} \underset{\sim}{A} \to 0$$

s a s.e.s. in $\underline{S}$. Then there is a covering T of R such that -

$$0 \to \underset{\sim}{C \otimes T} \xrightarrow{\underset{\sim}{\varphi \otimes T}} \underset{\sim}{H \otimes T} \xrightarrow{\underset{\sim}{\varrho \otimes T}} \underset{\sim}{A \otimes T} \to 0$$

is a s.e.s. of presheaves on $(\underline{A}')^{op}$, where $\underline{A}' = \underline{A}(T)$, $\underset{\sim}{C \otimes T} = \underline{A}'(C \otimes T,-)$, etc.

Proof. As shown in the proof of Proposition 15.6, there is a covering T of R such that -

$$0 \to \underset{\sim}{C}\sigma \to \underset{\sim}{H}\sigma \to \underset{\sim}{A}\sigma \to 0$$

is a s.e.s. of presheaves on $(\underline{A}')^{op}$, where σ: $\underline{A}' \to \underline{A}$ is the restriction functor (i.e., view T-algebras as R-algebras.) But, as was mentioned above, $\underset{\sim}{C}\sigma \approx \underline{A}'(C \otimes T,-)$, etc., and the assertion follows easily.

We denote the remainder of this section to a careful analysis of some special extensions of sheaves which will play a crucial role in the proof of our main theorem.

Lemma 15.8. Given a sequence -

$$A \xrightarrow{Q} H \xrightarrow{\varphi} RZ$$

in $\underline{H}$, where RZ is the group algebra of the infinite cyclic group $Z = \langle t \rangle$ with its usual Hopf algebra structure. Assume that $\varphi Q = \varepsilon_A$. Then H is a Z-graded R-algebra

$$H = \sum_{-\infty}^{+\infty} \oplus H_n$$

such that each H_n is a sub-coalgebra of H, $\varphi(z) = \varepsilon_H(z)t^n$ for z in H_n, and Q maps A into H_0. Finally, H_1 is an A-object in the category $\underline{C}$, with right A-module structure defined by $za = zQ(a)$ for z in H_1, a in A.

Proof. It is easily verified that the map α_H: $H \to H \otimes RZ$, defined to be the composite -

$$H \xrightarrow{\Delta_H} H \otimes H \xrightarrow{H \otimes \varphi} H \otimes RZ$$

gives to H the structure of an RZ-object in the category $\underline{A}$. Thus, setting $H_n = \{z \text{ in } H / \alpha_H(z) = z \otimes t^n\}$, we have from Chapter I, Proposition 4.17 that $H = \sum_{-\infty}^{+\infty} \oplus H_n$ is a Z-grading on H, with $H_m H_n \subseteq H_{m+n}$. In particular, H_0 is an R-subalgebra of H. Moreover, since α_H is an $\underline{H}$-map, a routine computation shows that each H_n is a sub-coalgebra of H, whence H_0 is a Hopf subalgebra of H.

If z is in H_n, we apply $\varepsilon_H \otimes 1$ to the equation -

$$\alpha_H(z) = \sum_{(z)} z_{(1)} \otimes \varphi(z_{(2)}) = z \otimes t^n$$

to obtain that $\varphi(z) = \varepsilon_H(z)t^n$ (the summation convention used above was introduced in Section 7.) Since $\varphi\varrho = \varepsilon_A$, it is then clear that $\varrho(A) \subseteq H_0$, and thus ϱ may be viewed as an $\underline{H}$-map from A to H_0. Finally, the $\underline{C}$-map α_1: $H_1 \otimes A \to H_1$ defined to be the composition -

$$H_1 \otimes A \xrightarrow{H_1 \otimes \varrho} H_1 \otimes H_0 \to H_1$$

(the unlabeled arrow denoting multiplication in H) gives to H_1 the structure of an A-object in the category $\underline{C}$, because ϱ: $A \to H_0$ is an $\underline{H}$-map. This completes the proof of the lemma.

Lemma 15.9. Let the sequence -

$$A \xrightarrow{\varrho} H \xrightarrow{\varphi} RZ$$

be as in Lemma 15.8. Assume that A is in $|\underline{H}_0|$ and -

$$0 \to \underset{\sim}{RZ} \xrightarrow{\underset{\sim}{\varphi}} \underset{\sim}{H} \xrightarrow{\underset{\sim}{\varrho}} \underset{\sim}{A} \to 0$$

is a s.e.s. of presheaves. Then H_1 is a finitely generated faithful projective R-module and ϱ: $A \to H_0$ and the map γ_1: $H_1 \otimes A \to H_1 \otimes H_1$ defined by -

$$\gamma_1(z \otimes a) = \Delta_H(z)(1 \otimes a) = \sum_{(z)} z_{(1)} \otimes z_{(2)}a \qquad (z \text{ in } H_1,\ a \text{ in } A)$$

are both isomorphisms.

Proof. Applying Theorem 14.2 and the Yoneda Lemma, we see that we may assume without loss of generality that $H = A \otimes RZ = AZ$ as R-algebras, with ϱ: $A \to AZ$ the natural map and φ: $AZ \to RZ$ satisfying the equation -

$$\varphi(at^n) = \varepsilon_A(a)t^n \qquad (a \text{ in } A)$$

A more careful analysis of the situation discussed in Theorem 14.2 yields easily that the identity H = AZ may be assumed to hold even as RZ-objects, whence $H_n = At^n \subseteq AZ$.[10] In particular, ϱ: $A \to H_0 = A\circ 1$ is an isomorphism. Moreover, since A is in $|H_0|$, $H_1 = At$ is a finitely generated faithful projective R-module.

Now, the fact that $H_1 = At$ is a sub-coalgebra of H guarantees that $\Delta_H(t)$ is in $(A \otimes A)(t \otimes t) \subseteq H \otimes H$. Thus we may write $\Delta_H(t) = u(t \otimes t)$, with u in $A \otimes A$. It is then easily verified that the mapping γ_1: $H_1 \otimes A \to H_1 \otimes H_1$ satisfies the equation -

$$\gamma_1(at \otimes b) = (\sum_{(a)} a_{(1)} \otimes a_{(2)} b) u(t \otimes t) \qquad (a,b \text{ in } A)$$

However since t is a unit of H and Δ_H is an $\underline{A}$-map, it follows that u is a unit of $A \otimes A$. A routine computation using the antipode of A then establishes that γ_1 is an isomorphism, completing the proof of the lemma.

Theorem 15.10. Let the sequence -

$$A \xrightarrow{\varrho} H \xrightarrow{\varphi} RZ$$

be as in Lemma 15.8, with A in $|\underline{H}_0|$. Assume that -

$$0 \to \underset{\sim}{RZ} \xrightarrow{\varphi} \underset{\sim}{H} \xrightarrow{\varrho} \underset{\sim}{A} \to 0$$

is a s.e.s. of sheaves. Then $\varrho: A \to H_0$ is an isomorphism, and H_1 is a finitely generated faithful projective R-module and a Galois A-object in the category $\underline{C}$ (we refer the read to Section 4 for a discussion of these.)

Proof. Remark 15.7 guarantees the existence of a covering T of R such that the sequence -

$$0 \to \underset{\sim}{TZ} \xrightarrow{\varphi \otimes T} \underset{\sim}{H \otimes T} \xrightarrow{\varrho \otimes T} \underset{\sim}{A \otimes T} \to 0$$

is a s.e.s. of presheaves on $\underline{A}(T)^{op}$. We may then apply the preceding lemmas, with T playing the role of R, observing first that the Z-grading on $H \otimes T$ arising from Lemma 15.8 clearly satisfies the condition $(H \otimes T)_n = H_n \otimes T$. By Lemma 15.9, the $\underline{H}_0(T)$ -map $\varrho \otimes T : A \otimes T \to H_0 \otimes T$ is an isomorphism, whence ϱ is an isomorphism because T is a faithfully flat R-module. A similar argument shows that $\gamma_1: H_1 \otimes A \to H_1 \otimes H_1$ is an isomorphism. Finally, $H_1 \otimes T$ is a finitely generated faithful projective T-module, by Lemma 15.9, from which we obtain via [10, Chapitre I, Proposition 12, p. 53] that H_1 is a finitely generated faithful projective R-module. In particular, H_1 is a faithfully flat R-module, and is thus clearly a cofaithful object of $\underline{C}$. We may then conclude that H_1 is a Galois A-object in $\underline{C}$, and the proof of the theorem is complete.

16. The Isomorphism

We shall denote by U the abelian presheaf on $\underline{A}^{op}$ defined as follows: If S is in $|\underline{A}|$, hen U(S) is the multiplicative abelian group of invertible elements of S. One sees easiy that, in the notation of Section 15, $U = \underset{\sim}{RZ} = \underline{A}(RZ,-)$, and is hence a sheaf by Propoition 15.4.

Now let A be in $|H_0|$, and S be an A-object such that the natural map $R \to S$ is oneo-one. It is trivially verified that $\Omega = Hom_R(A^*,S)$ is an R-algebra with unit ε_{A^*} and ultiplication defined by the formula -

16.1) $$(\varphi\circ\psi)(u) = \sum_{(u)} \varphi(u_{(1)})\psi(u_{(2)}) \qquad (u \text{ in } A^*;\ \varphi,\psi \text{ in } \Omega)$$

f s is in U(S), we define φ_x in Ω by the formula -

$$\varphi_x(u) = x^{-1}u(x) \qquad (u \text{ in } A^*)$$

(x) being as in (7.2). Note that $\varphi_1 = \varepsilon_{A^*}$; furthermore, an easy computation using (7.2) stablishes that -

16.2) $$u(xy) = \sum_{(u)} u_{(1)}(x)u_{(2)}(y) \qquad (x,y \text{ in } S;\ u \text{ in } A^*)$$

rom which we obtain immediately that $\varphi_{xy} = \varphi_x\circ\varphi_y$ for x,y in U(S). Thus -

16.3) The mapping $x \to \varphi_x$ is a homomorphism of U(S) into $U(\Omega)$.

Next we define a subset V(S) of U(S) by the condition

16.4) $$V(S) = \{x \text{ in } U(S) \mid x^{-1}u(x) \text{ is in } R \text{ for all } u \text{ in } A^*\}.$$

f u,v are in A*, then for x in V(S) we have that $\varphi_x(uv) = x^{-1}uv(x) = x^{-1}u\{x(x^{-1}v(x))\} = x^{-1}u(x))(x^{-1}v(x)) = \varphi_x(u)\varphi_x(v)$, since $x^{-1}v(x)$ is in R. Furthermore -

$$1(x) = \sum_{(x)} x_{(1)}\langle 1,x_{(2)}\rangle = \sum_{(x)} x_{(1)}\varepsilon_A(x_{(2)}) = x$$

y (7.2), whence $\varphi_x(1) = 1$. Thus, if x is in V(S), then φ_x is in $\underline{A}(A^*,R)$, which is in natural way a subgroup of $U(\Omega)$. It is then clear that an element x of U(S) is in V(S) f and only if φ_x is in $\underline{A}(A^*,R)$. (16.3) then guarantees that V(S) is a subgroup of U(S), and the mapping $x \to \varphi_x$ is a homomorphism of V(S) into $\underline{A}(A^*,R)$. Finally, it is obvious that $U(R) \subseteq V(S)$, and (7.2) and an easy computation show that $\varphi_x = \varepsilon_{A^*}$ if x is in U(R).

Lemma 16.5. Let A be in $|H_0|$, and S be a Galois A-object. Then the sequence -

$$0 \to U(R) \to V(S) \to \underline{A}(A^*,R)$$

is exact, the maps being the inclusion and $x \to \varphi_x$, respectively.

Proof. Assume that s is in V(S) and $\varphi_x = \varepsilon_{A^*}$. Then, for all u in A^*, $u(x) = x\varepsilon_{A^*}(u)$ whence x is in $S^{A^*} = R$, by (7.5) and Theorem 7.6(a). The remainder of the lemma has been established in the preceding discussion.

Our next task is to derive conditions under which the map $V(S) \to \underline{A}(A^*,R)$ of Lemma 16.5 is onto. We shall make use of the right A^*-module structure on $S^* = \mathrm{Hom}_R(S,R)$ (S an A-object) which is induced in the usual way by the left A^*-module structure on S of (7.2). It is easy to see that this coincides with the A^*-module structure arising from the composite map -

$$S^* \otimes A^* \to (S \otimes A)^* \xrightarrow{\alpha^*_S} S^*$$

with $\alpha_S: S \to S \otimes A$ as in Section 7.

Lemma 16.6. Let S be a Galois A-object, with A in $|H_0|$. If $S^* \approx A^*$ as right A^*-modules, then the map $V(S) \to \underline{A}(A^*,R)$ is onto.

Proof. By hypothesis, there exists z in S^* such that the map $A^* \to S^*$ defined by $u \to zu$ is an isomorphism. Now, given φ in $\underline{A}(A^*,R)$, define $f: S^* \to R$ by the formula $f(zu) = \varphi(u)$. f is clearly a homomorphism of R-modules; hence, since S is a finitely generated projective R-module by Theorem 7.6(a), there exists a unique x in S such that $f = \langle -,x\rangle$, with $\langle\ \rangle: S^* \otimes S \to R$ the duality pairing. For any u,v in A^*, we then have $\langle zu,v(x)\rangle=\langle zuv,x\rangle=\varphi(uv)=\varphi(u)\varphi(v)=\langle zu,x\rangle\varphi(v)=\langle zu,x\varphi(v)\rangle$, whence our hypotheses on S^* guarantee that $v(x) = x\varphi(v)$ for all v in A^*. Therefore, if x is in U(S), we may conclude that $x^{-1}v(x) = \varphi(v)$ is in R for all v in A^*, whereby x is in V(S) and $\varphi = \varphi_x$. Thus, to complete the proof of the lemma, we need only show that x is in U(S).

To this end, let us first consider the special case in which S = A. Then $S^* = A^* = zA^*$, and so we can find u in A^* with zu = 1, in which case $\varepsilon_A(x) = \langle 1,x\rangle = \langle zu,x\rangle = \varphi(u)$ is in U(R), since u is in $U(A^*)$. Now, $\underline{A}(A^*,R) \subseteq \mathrm{Hom}_R(A^*,R) = A^{**} \approx A$, the latter eing the usual natural isomorphism which arises from the fact that A is a finitely enerated projective R-module. Indeed, the definition of the Hopf algebra structure on *, etc., guarantees that $\underline{A}(A^*,R)$ is even a subgroup of U(A). The equation -

$$x\varphi(u) = u(x) = \sum_{(x)} x_{(1)} \langle u, x_{(2)} \rangle$$

hown above to hold for all u in A*, then yields that $x \otimes \varphi = \Delta_A(x)$ in $A \otimes A \approx \mathrm{Hom}_R(A^*,A)$. pplying $\varepsilon_A \otimes 1_A$ to both sides of this equation, we obtain that $x = \varepsilon_A(x)\varphi$ is in U(A), ompleting the proof for the case in which S = A.

Now let S be an arbitrary Galois A-object. Then $S \otimes A$ is in $|H_0(S)|$, and $S \otimes S$ is, n the obvious way, an $S \otimes A$-object (S acting on the left-most factor.) In fact, by Defini- ion 7.3(b), $S \otimes S \approx S \otimes A$ as $S \otimes A$-objects. We may then apply the preceding discussion, ith $1 \otimes x$ in $S \otimes S$, $1 \otimes \varphi$: $S \otimes A^* \approx \mathrm{Hom}_S(S \otimes A, S) \to S$, and $1 \otimes z$ in $S \otimes S^* \approx \mathrm{Hom}_S(S \otimes S, S)$ laying the roles of x, φ, z, respectively, to obtain that $1 \otimes x$ is in $U(S \otimes S)$. Since S s a faithfully flat R-module by Definition 7.3, it follows easily that x is in U(S), ompleting the proof of the lemma.

Next we consider the way in which V(S) varies with A. Recall that, as discussed in ections 2 and 4, a map h: B → A in |H| gives rise to a functor $\tilde{h}$: $\underline{A}^A \to \underline{A}^B$. (Recall that ^{A} is the category of A-objects and maps of such, as defined in Section 1. The change in ariance arises from the fact that we are dealing with cogroups in $\underline{A}$ rather than with roups in $\underline{A}$.) If S is an A-object, then $\tilde{h}(S)$ is given by the equalizer diagram -

$$\tilde{h}(S) \to S \otimes B \underset{\omega}{\overset{\beta}{\rightrightarrows}} S \otimes A \otimes B$$

ith β and ω satisfying the formulae -

$$\beta(x \otimes b) = \sum_{(x)} x_{(1)} \otimes x_{(2)} \otimes b \qquad\qquad (x \text{ in } S;\ b \text{ in } B)$$

$$\omega(x \otimes b) = \sum_{(b)} x \otimes h(b_{(1)}) \otimes b_{(2)}$$

hat is, $\tilde{h}(S) = \{y \text{ in } S \otimes B \mid \beta(y) = \omega(y)\}$. $\alpha_{\tilde{h}(S)}: \tilde{h}(S) \to \tilde{h}(S) \otimes B$ is that unique map endering the diagram below commutative -

$$\begin{array}{ccc}
\tilde{h}(S) & \xrightarrow{\alpha_{\tilde{h}(S)}} & \tilde{h}(S) \otimes B \\
\downarrow & & \downarrow \\
S \otimes B & \xrightarrow{S \otimes \Delta_B} & S \otimes B \otimes B
\end{array}$$

For a discussion of these facts we refer the reader to Theorem 2.9 .A routine computati then shows that the diagram below also commutes -

(16.7)
$$\begin{array}{ccc}
\tilde{h}(S) & \xrightarrow{\alpha_{\tilde{h}(S)}} & \tilde{h}(S) \otimes B \\
\downarrow \theta_S & & \downarrow \theta_S \otimes h \\
S & \xrightarrow{\alpha_S} & S \otimes A
\end{array}$$

with θ_S the composite -

$$\tilde{h}(S) \to S \otimes B \xrightarrow{S \otimes \varepsilon_B} S$$

This yields easily, via (7.2), the formula -

(16.8) $u(\theta_S(x)) = \theta_S\{h^*(u)(x)\}$ for all x in $\tilde{h}(S)$, u in A^* with $h^*: A^* \to B^*$ the dual of h.

<u>Lemma 16.9.</u> (a) Let $h: B \to A$ be an $\underline{H}_0$-map, and S be a Galois A-object. Then the ma $\theta_S: \tilde{h}(S) \to S$ of (16.7) maps $V(\tilde{h}(S))$ into $V(S)$, and the diagram below commutes -

$$\begin{array}{ccccc}
U(R) & \longrightarrow & V(\tilde{h}(S)) & \longrightarrow & \underline{A}(B^*,R) \\
\| & & \downarrow & & \downarrow \underline{A}(h^*,R) \\
U(R) & \longrightarrow & V(S) & \longrightarrow & \underline{A}(A^*,R)
\end{array}$$

the horizontal maps being as in Lemma 16.5, and the middle vertical map being the restriction of θ_S.

(b) Let S_i be Galois A_i-objects (A_i in $|\underline{H}_0|$, $i = 1,2$), in which case $S_1 \otimes S_2$ is a Galois $A_1 \otimes A_2$ -object by Proposition 3.2 . Then the diagram below commutes -

$$\begin{array}{ccccc}
U(R)\times U(R) & \longrightarrow & V(S_1)\times V(S_2) & \longrightarrow & \underline{A}(A_1^*,R)\times A(A_2^*,R) \\
\downarrow & & \downarrow & & \| \\
U(R) & \longrightarrow & V(S_1 \otimes S_2) & \longrightarrow & \underline{A}((A_1 \otimes A_2)^*,R)
\end{array}$$

with the horizontal maps as in Lemma 16.5 and the vertical ones as follows: $(x,y) \to xy$ for x,y in $U(R)$; $(x_1,x_2) \to x_1 \otimes x_2$ for x_i in $V(S_i)$; and $(\varphi_1,\varphi_2) \to \varphi_1 \otimes \varphi_2$ for φ_i in $\underline{A}(A_i^*,R)$, $i = 1,2$ (note that this last map is an isomorphism).

Proof. (a) is an easy consequence of (16.8) and a routine diagram chase, and (b) follows from the definitions.

Having established these preliminaries, we now introduce the appropriate setting for our work. Let A be in $|H_0|$, and S be a Galois A-object. Then, for T in $|\underline{A}|$, $A \otimes T$ is in $|H_0(T)|$ and $S \otimes T$ is a Galois $A \otimes T$ -object. Thus we may define the abelian groups $V(S \otimes T)$ as in (16.4), with T, $A \otimes T$, $S \otimes T$ playing the roles of R, A, and S, respectively. We therefore obtain easily an abelian presheaf V_S on $\underline{A}^{op}$, defined by the conditions below -

(16.10) $\qquad V_S(T) = V(S \otimes T) \quad$ for T in $|\underline{A}|$

$\qquad V_S(f)\colon V_S(T) \to V_S(T') \quad$ for $f\colon T \to T'$ in $\underline{A}$ is the restriction to $V(S \otimes T)$ of the map $S \otimes f\colon S \otimes T \to S \otimes T'$.

In the remainder of this section we shall make repeated (and often implicit) use of standard identities such as the one below -

(16.11) $\qquad \underline{A}(A^*,T) \approx \underline{A}(T)(A^* \otimes T,T) \approx \underline{A}(T)(\mathrm{Hom}_T(A \otimes T,T),T) \quad$ for T in $|\underline{A}|$.

<u>Proposition 16.12.</u> Let S be a Galois A-object, with A in $|H_0|$. Then V_S is a sheaf, and the sequence -

$$\xi_S\colon 0 \to U \to V_S \xrightarrow{\pi_S} \underset{\sim}{A^*} \to 0$$

is a s.e.s. in $\underline{S}$; where, for T in $|A|$, the maps $U(T) \to V(S \otimes T)$ and $\pi_S(T)\colon V(S \otimes T) \to \underline{A}(A^*,T) \approx \underline{A}(T)(\mathrm{Hom}_T(A \otimes T,T),T)$ are those of Lemma 16.5, T playing the role of R.

Proof. U is a sheaf, as remarked at the beginning of this section, and A^* is a sheaf by Proposition 15.4. As for V_S, let T be a covering of R; i.e., of the form (15.1). Consider the commutative diagram -

$$\begin{array}{ccccc}
0 \to V(S) & \xrightarrow{d} & V(S \otimes T) & \xrightarrow{d^0 - d^1} & V(S \otimes T \otimes T) \\
\downarrow & & \downarrow & & \downarrow \\
0 \to U(S) & \xrightarrow{d} & U(S \otimes T) & \xrightarrow{d^0 - d^1} & U(S \otimes T \otimes T)
\end{array}$$

$d: S \to S \otimes T$ and $d^i: S \otimes T \to S \otimes T \otimes T$ $(i = 1,2)$ being as in (15.2), and the vertical maps being the inclusions. Since U is a sheaf, the lower sequence is exact. Thus the upper sequence is exact at $V(S)$, and x in $V(S \otimes T)$ has the property that $d^0(x) = d^1(x)$ if and only if $x = y \otimes 1$ for some y in $U(S)$. If u is in A^*, then $v = u \otimes 1$ is in $A^* \otimes T \approx \mathrm{Hom}_T(A \otimes T,T)$, whence $(y^{-1}u(y)) \otimes 1 = x^{-1}v(x)$ is in T. Since T is a faithfully flat R-module, we obtain that $y^{-1}u(y)$ is in R, whereupon y is in $V(S)$. We may then conclude that the upper sequence above is exact; this is simply the sequence -

$$0 \to V_S(R) \xrightarrow{V_S(d)} V_S(T) \xrightarrow{V_S(d^0) - V_S(d^1)} V_S(T \otimes T)$$

If $d: T_0 \to T_0 \otimes T$ is any covering, then the proof that the sequence -

$$0 \to V_S(T_0) \xrightarrow{V_S(d)} V_S(T_0 \otimes T) \xrightarrow{V_S(d^0) - V_S(d^1)} V_S(T_0 \otimes T \otimes T)$$

is exact is entirely similar to the above argument replacing R by T_0, S by $S \otimes T_0$, etc. That V_S is a sheaf then follows from (15.2).

We turn now to the exactness of the sequence ξ_S. Routine computations establish that the maps π_S and $U \to V_S$ are sheaf homomorphisms. Moreover, if T_0 is in $|A|$, we have from Lemma 16.5 (with T_0 playing the role of R) that the following sequence of abelian groups is exact -

$$0 \to U(T_0) \to V_S(T_0) \xrightarrow{\pi_S(T_0)} \underline{A}(A^*,T_0)$$

Furthermore, by Lemma 16.3 below, S^* is an invertible right A^*-module, and thus we may utilize the methods of [13, Section 5] to obtain a covering T of R such that $S^* \otimes T \approx A^* \otimes T$ as right $A^* \otimes T$-modules. Then $T_0 \to T_0 \otimes T = T'$ is a covering of T_0, and $\mathrm{Hom}_{T'}(S \otimes T',T') \approx S^* \otimes T' \approx A^* \otimes T' \approx \mathrm{Hom}_T(A \otimes T',T')$ as right $\mathrm{Hom}_{T'}(A \otimes T',T')$-modules. We may then apply Lemma 16.6 (with T' playing the role of R) to conclude that $\pi_S(T')$ is onto, whence (15.3) yields easily that the sequence ξ_S is exact in $\underline{S}$. This completes the proof.

Lemma 16.13. Let A be in $|H_0|$, and S be a Galois A-object. Then S^* is an invertible right A^*-module (an invertible module is a projective module of rank one; see [10, Chapitre 2, §4].)

Proof. Since S, and hence S*, is a finitely generated projective R-module by Theorem 7.6(a), we may apply Proposition 4.8. to obtain that S* is a Galois A*-object in the category $\underline{C}$. Therefore the map $\gamma_{S^*}: S^* \otimes A^* \to S^* \otimes S^*$, where -

$$\gamma_{S^*}(z \otimes u) = \Delta_{S^*}(z)(1 \otimes u) \qquad (z \text{ in } S^*,\ u \text{ in } A^*)$$

is an isomorphism, whence $S^* \otimes A^* \approx S^* \otimes S^*$ as right A*-modules, A* acting on the rightmost factors. Since, by Theorem 9.3, S and hence S* possess direct summands isomorphic to R, the lemma follows from a routine direct sum argument and counting of ranks.

Our main result, stated below, involves the additive functor $X: \underline{H}^{op} \approx Ab(\underline{A}^{op}) \to Ab$, which is defined for arbitrary categories in Section 3 and interpreted for the category of commutative algebras in Section 4.

<u>Theorem 16.14.</u> Given A in $|\underline{H}_0|$, there exists an isomorphism -

$$j_A: X(A) \xrightarrow{\approx} \mathrm{Ext}^1_{\underline{S}}(\underset{\sim}{A^*}, U)$$

which is natural in A. If S is a Galois A-object, then $j_A(cl(S)) = cl(\xi_S)$, where -

$$\xi_S: 0 \to U \to V_S \xrightarrow{\pi_S} \underset{\sim}{A^*} \to 0$$

is the s.e.s. of sheaves of Proposition 16.12.

Most of the remainder of this section will be devoted to the proof of Theorem 16.14. We begin by observing that j_A is well-defined. For, if $f: S \to S'$ in $\underline{A}^A$, with S, S' Galois A-objects, then we obtain a commutative diagram -

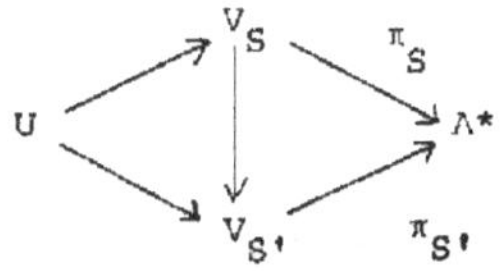

the vertical map being induced in the obvious way by f; thus $cl(\xi_S) = cl(\xi_{S'})$ in $\mathrm{Ext}^1_{\underline{S}}(\underset{\sim}{A^*}, U)$.

Next we show that j_A is natural in A. Let $h: B \to A$ in $|\underline{H}_0|$, and S be a Galois A-object. If T is in $|\underline{A}|$, an easy application of Corollary 2.17 yields that $\widetilde{h \otimes T}(S \otimes T) \approx \tilde{h}(S) \otimes T$ in $\underline{A}(T)^{A \otimes T}$, whence $V_{\tilde{h}(S)}(T) = V(\tilde{h}(S) \otimes T) \approx V\{\widetilde{h \otimes T}(S \otimes T)\}$. We may then replace R by T, etc., in Lemma 16.9(a) to obtain the commutative diagram -

$$\begin{array}{ccccc}
U(T) & \longrightarrow V_{\tilde{h}(S)}(T) & \xrightarrow{\pi_{\tilde{h}(S)}(T)} & \underline{A}(B^*,T) & \\
\| & \downarrow & & \downarrow & \underline{A}(h^*,T) \\
U(T) & \longrightarrow V_S(T) & \xrightarrow{\pi_S(T)} & \underline{A}(A^*,T) &
\end{array}$$

Thus we have a commutative diagram in $\underline{S}$ -

$$\begin{array}{ccccccccc}
0 & \longrightarrow & U & \longrightarrow & V_{\tilde{h}(S)} & \xrightarrow{\pi_{\tilde{h}(S)}} & \underset{\sim}{B^*} & \longrightarrow & 0 \\
 & & \| & & \downarrow & & \downarrow \underset{\sim}{h^*} & & \\
0 & \longrightarrow & U & \longrightarrow & V_S & \xrightarrow{\pi_S} & \underset{\sim}{A^*} & \longrightarrow & 0
\end{array}$$

Since $X(h)(cl(S)) = cl(\tilde{h}(S))$ in $X(B)$, it follows that the diagram below commutes -

$$\begin{array}{ccc}
X(A) & \xrightarrow{j_A} & \mathrm{Ext}^1_{\underline{S}}(\underset{\sim}{A^*},U) \\
\downarrow X(h) & & \downarrow \mathrm{Ext}^1_{\underline{S}}(\underset{\sim}{h^*},U) \\
X(B) & \xrightarrow{j_B} & \mathrm{Ext}^1_{\underline{S}}(\underset{\sim}{B^*},U)
\end{array}$$

establishing the naturality of j_A.

We turn now to the argument that j_A is a homomorphism of groups. Given A_1, A_2 in $\underline{\mathbb{H}}$ we show first that the diagram below commutes -

(16.15)
$$\begin{array}{ccc}
X(A_1)\times X(A_2) & \xrightarrow{j_{A_1}\times j_{A_2}} & \mathrm{Ext}^1_{\underline{S}}(\underset{\sim}{A_1^*},U)\times \mathrm{Ext}^1_{\underline{S}}(\underset{\sim}{A_2^*},U) \\
\downarrow & & \downarrow \\
X(A_1\otimes A_2) & \xrightarrow{j_{A_1\otimes A_2}} & \mathrm{Ext}^1_{\underline{S}}(\underset{\sim}{(A_1\otimes A_2)^*},U)
\end{array}$$

the left vertical map being that of Definition 3.7 $((cl(S_1),\ cl(S_2)) \to cl(S_1\otimes S_2))$ a[nd] the right vertical one being the map $(cl(\xi_1), cl(\xi_2)) \to cl(\xi)$; where, if -

$$\xi_i : 0 \to U \to E_i \to \underset{\sim}{A_i^*} \to 0$$

then ξ is given by the commutative diagram -

$$\begin{array}{cccccccccc}
 & 0 & \to & U\times U & \to & E_1\times E_2 & \to & \underset{\sim}{A_1^*}\times \underset{\sim}{A_2^*} & \to & 0 \\
 & & & \downarrow & & \downarrow & & \| & & \\
\xi : & 0 & \to & U & \to & E & \to & \underset{\sim}{(A_1\otimes A_2)^*} & \to & 0
\end{array}$$

The left-most vertical map in this diagram is the multiplication map, and the right-most one, an isomorphism, arises from the chain of identities -

$$\underset{\sim}{A}_1^* \times \underset{\sim}{A}_2^* = \underline{A}(A_1^*,-) \times \underline{A}(A_2^*,-) \approx \underline{A}(A_1^* \otimes A_2^*,-) \approx \underline{A}((A_1 \otimes A_2)^*,-) = \underset{\sim}{(A_1 \otimes A_2)}^*$$

Now let S_i be a Galois A_i-object $(i = 1,2)$ and T be in $|\underline{A}|$. Then, upon replacing R by T, etc., in Lemma 16.9(b), we obtain a commutative diagram -

$$\begin{array}{ccccc}
U(T)\times U(T) & \longrightarrow & V_{S_1}(T)\times V_{S_2}(T) & \xrightarrow{\pi_{S_1}(T)\times\pi_{S_2}(T)} & \underline{A}(A_1^*,T)\times\underline{A}(A_2^*,T) \\
\downarrow & & \downarrow & & \downarrow \\
U(T) & \longrightarrow & V_{S_1\otimes S_2}(T) & \xrightarrow{\pi_{S_1\otimes S_2}(T)} & \underline{A}((A_1\otimes A_2)^*,T)
\end{array}$$

the left-most vertical map being multiplication. This yields the commutative diagram below in $\underline{S}$ -

$$\begin{array}{ccccccccc}
0 & \longrightarrow & U\times U & \longrightarrow & V_{S_1}\times V_{S_2} & \xrightarrow{\pi_{S_1}\times\pi_{S_2}} & \underset{\sim}{A}_1^*\times\underset{\sim}{A}_2^* & \longrightarrow & 0 \\
 & & \downarrow & & \downarrow & & \| & & \\
0 & \longrightarrow & U & \longrightarrow & V_{S_1\otimes S_2} & \xrightarrow{\pi_{S_1\otimes S_2}} & \underset{\sim}{(A_1\otimes A_2)}^* & \longrightarrow & 0
\end{array}$$

The commutativity of (16.15) then follows immediately from the definition of the various maps involved.

Now, if A is in $|\underline{H}_0|$, we obtain the commutative diagram -

$$\begin{array}{ccc}
X(A)\times X(A) & \xrightarrow{j_A\times j_A} & \mathrm{Ext}^1_{\underline{S}}(\underset{\sim}{A}^*,U)\times\mathrm{Ext}^1_{\underline{S}}(\underset{\sim}{A}^*,U) \\
\downarrow & & \downarrow \\
X(A\otimes A) & \xrightarrow{j_{A\otimes A}} & \mathrm{Ext}^1_{\underline{S}}(\underset{\sim}{(A\otimes A)}^*,U) \\
\downarrow X(\Delta_A) & & \downarrow \mathrm{Ext}^1_{\underline{S}}(\underset{\sim}{\Delta}_A^*,U) \\
X(A) & \xrightarrow{j_A} & \mathrm{Ext}^1_{\underline{S}}(\underset{\sim}{A}^*,U)
\end{array}$$

the upper and lower squares being special cases of (16.15) and naturality, respectively. Since the composite vertical maps give by definition the addition in X(A) and $\mathrm{Ext}^1_{\underline{S}}(\underset{\sim}{A}^*,U)$, we conclude that j_A is a homomorphism of abelian groups.

Our next task is to prove that j_A is one-to-one. First, however, we record the following useful remark.

<u>Remark 16.16.</u> Let S be a Galois A-object (A in $|\underline{H}_O|$), T be in $|\underline{A}|$, x be in $V_S(T)$, and set $\varphi_x = \pi_S(T)(x)$ in $\underline{A}(A^*,T)$. Then $\langle zu,x\rangle_T = \langle z,x\rangle_T\varphi_x(u)$ for all u in A* and z in S* with $\langle\ \rangle_T\colon S^*\otimes(S\otimes T)\to T$ the duality pairing.

Proof. Since $x^{-1}(u\otimes 1)(x) = \varphi_x(u)$ is in T by definition of V_S, we have that $\langle zu,x\rangle_T = \langle z,(u\otimes 1)(x)\rangle_T = \langle z,x\rangle_T(x^{-1}(u\otimes 1)(x)) = \langle z,x\rangle_T\varphi_x(u)$, completing the proof.

Suppose now that $j_A(\mathrm{cl}(S)) = 0$ for some Galois A-object S, with A in $|\underline{H}_O|$. Then the sheaf extension -

$$\xi_S\colon 0\to U\to V_S\xrightarrow{\pi_S}\underset{\sim}{A^*}\to 0$$

splits; i.e., there is a homomorphism $\sigma\colon A^*\to V_S$ of abelian sheaves such that $\pi_S\sigma$ is the identity map of $\underset{\sim}{A^*}$. Letting 1_{A^*} in $\underline{A}(A^*,A^*)$ be the identity map of A*, we set $x=\sigma(A^*)(1_{A^*})$ in $V_S(A^*) = V(S\otimes A^*)$, in which case $\pi_S(A^*)(x) = 1_{A^*}$. Setting T = A* in Remark 16.16, we then obtain that $\langle zu,x\rangle_{A^*} = \langle z,x\rangle_{A^*}u$ for all u in A*, z in S*, whence the mapping $f = \langle -,x\rangle_{A^*}\colon S^*\to A^*$ is a homomorphism of right A*-modules.

Since, for T in $|\underline{A}|$, $\sigma(T)\colon\underline{A}(A^*,T)\to V_S(T)$ is a homomorphism of abelian groups, we obtain the commutative diagram below -

$$\begin{array}{ccc} \underline{A}(A^*,T)\times\underline{A}(A^*,T) & \xrightarrow{\quad\sigma(T)\otimes\sigma(T)\quad} & V_S(T)\times V_S(T)\\ \Vert & & \\ \underline{A}(A^*\otimes A^*,T) & & \Big\downarrow \\ \Big\downarrow{\scriptstyle \underline{A}(\Delta_{A^*},T)} & & \\ \underline{A}(A^*,T) & \xrightarrow{\quad\sigma(T)\quad} & V_S(T)\end{array}$$

where the equality and unlabeled arrow denote the natural isomorphism and multiplication map, respectively. Of course, if $\varphi\colon A^*\otimes A^*\to T$ in $\underline{A}(A^*\otimes A^*,T)$, then the corresponding element in $\underline{A}(A^*,T)\times\underline{A}(A^*,T)$ is (φ_1,φ_2), where $\varphi_1(u) = \varphi(u\otimes 1)$ and $\varphi_2(u) = \varphi(1\otimes u)$ for u in A*.

Now let $\psi_1,\psi_2\colon A^*\to A^*\otimes A^*$ be defined by $\psi_1(u) = u\otimes 1$, $\psi_2(u) = 1\otimes u$ for u in A*. If we trace the element (ψ_1,ψ_2) through the above diagram, with $T = A^*\otimes A^*$, we see that

$$\sigma(A^*\otimes A^*)(\Delta_{A^*}) = [\sigma(A^*\otimes A^*)(\psi_1)][\sigma(A^*\otimes A^*)(\psi_2)]$$

in $V_S(A^* \otimes A^*)$. This, by naturality of σ and the definition of x, reduces immediately to the equation -

$$(1 \otimes \Delta_{A^*})(x) = [(1 \otimes \psi_1)(x)][(1 \otimes \psi_2)(x)]$$

in $V_S(A^* \otimes A^*) = V(S \otimes A^* \otimes A^*)$. Application of $\langle z,-\rangle_{A^* \otimes A^*}$ (with z in S^*) to both sides of this equation then yields easily that -

$$\Delta_{A^*}(f(z)) = \sum_{(z)} f(z_{(1)}) \otimes f(z_{(2)}) \qquad (z \text{ in } S^*)$$

That is, the map $f\colon S^* \to A^*$ preserves the coalgebra multiplication. A similar (but easier) argument shows that f preserves the coalgebra augmentation, and is therefore a map of coalgebras. Since f also preserves the right A^*-module structure, we obtain that f is a map of Galois A^*-objects in the category $\underline{C}$, and thus an isomorphism by Theorem 1.12 . Dualizing, we conclude that $S \approx A$ as Galois A-objects, whence $cl(S) = 0$ in $X(A)$ and j_A is one-to-one.

Finally, we show that j_A is onto. Let $cl(\xi)$ be in $\mathrm{Ext}^1_{\underline{S}}(\underset{\sim}{A^*},U)$, with -

$$\xi\colon 0 \to U \to E \to \underset{\sim}{A^*} \to 0$$

a s.e.s. in $\underline{S}$. Then, by Proposition 15.6, there exists a sequence -

$$A^* \xrightarrow{\ \rho\ } H \xrightarrow{\ \psi\ } R\mathbb{Z}$$

in $\underline{H}$ and an isomorphism $\zeta\colon E \overset{\approx}{\to} H$ of sheaves such that the diagram below commutes -

(16.17) 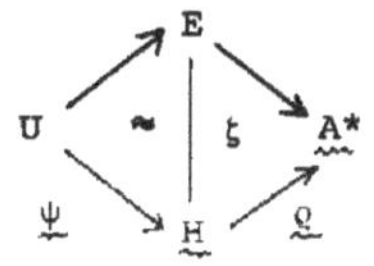

Furthermore, by Theorem 15.10, H is a $\mathbb{Z}$-graded R-algebra -

$$H = \sum_{-\infty}^{+\infty} \oplus H_n$$

such that each H_n is a sub-coalgebra of H, $\psi(z) = \varepsilon_H(z)t^n$ for z in H_n (where we write the group Z multiplicatively with generator t), and ρ maps A^* isomorphically onto H_0. Finally, by the same theorem, H_1 is a finitely generated projective R-module and a Galois A^*-object in the category $\underline{C}$ of cocommutative R-coalgebras, with right A^*-module structure defined by $zu = z\varrho(u)$ for z in H_1, u in A^*.

Dualizing and using Proposition 4.8, we then see that $S = H_1^*$ is a Galois A-object, with $\alpha_S : S \to S \otimes A$ the composite -

$$S = H_1^* \to (H_1 \otimes A)^* \approx S \otimes A$$

the unlabeled arrow denoting the transpose of the map $H_1 \otimes A^* \to H_1^*$ which gives the A^*-module structure on H_1. We claim that $j_A(\mathrm{cl}(S)) = \mathrm{cl}(\xi)$ in $\mathrm{Ext}^1_{\underline{S}}(\underset{\sim}{A^*}, U)$. To this end, we define a sheaf homomorphism $\theta : \underset{\sim}{H^*} \to V_S$ such that the diagram below commutes -

(16.18)
$$\begin{array}{ccccc} \underset{\sim}{R(Z)} & \xrightarrow{\psi} & \underset{\sim}{H} & \xrightarrow{\varrho} & \underset{\sim}{A^*} \\ \| & & \downarrow{\scriptstyle\theta} & & \| \\ U & \longrightarrow & V_S & \xrightarrow{\pi_S} & \underset{\sim}{A^*} \end{array}$$

Given T in $|\underline{A}|$ and f in $\underline{A}(H,T)$, the fact that S is a finitely generated projective R-module yields the existence of a unique element x_f in $S \otimes T$ such that -

$$f(z) = \langle z, x_f \rangle_T$$

for all z in $H_1 = S^*$, with $\langle\ \rangle_T : S^* \otimes (S \otimes T) \to T$ the duality pairing. If also g is in $\underline{A}(H,T)$ and $f+g$ denotes the sum in $\underline{A}(H,T)$ (i.e., $f+g$ is the composite -

$$H \xrightarrow{\Delta_H} H \otimes H \xrightarrow{f \otimes g} T \otimes T \longrightarrow T$$

the unlabeled arrow denoting the multiplication map), then the fact that $S^* = H_1$ is a sub-coalgebra of H guarantees via routine computation that -

(16.19)
$$x_{f+g} = x_f x_g$$

In particular, if $g = -f$, then $f+g = \varepsilon_H$, in which case -

$$\langle z, x_f x_g \rangle_T = \langle z, x_{f+g} \rangle_T = \varepsilon_H(z) = \varepsilon_{S^*}(z) = \langle z, 1 \otimes 1 \rangle_T$$

for all z in S^*, by (16.19). This means that $x_f x_g = 1 \otimes 1$ in $S \otimes T$, and so x_f is in $U(S \otimes T)$.

Suppose that f is in $\underline{A}(H,T)$ and u is in A^*. Then, for all z in S^*, we have from the definition of the A^*-module structure on S^* that $\langle z, (u \otimes 1)(x_f) \rangle_T = \langle zu, x_f \rangle_T =$ $\langle z\varrho(u), x_f \rangle_T = f(z\varrho(u)) = f(z) f(\varrho(u)) = \langle z, x_f f(\varrho(u)) \rangle_T$, whence -

(16.20) $$(u \otimes 1)(x_f) = x_f f(\varrho(u))$$

for all u in A*, f in $\underline{A}(H,T)$) It is then clear that $x_f^{-1}v(x_f)$ is in T for all v in $A^* \otimes T$, hence x_f is in $V(S \otimes T) = V_S(T)$. Moreover, if $\pi_S(T)(x_f) = \varphi$ in $\underline{A}(A^*,T)$, then $\varphi(u) = x_f^{-1}(u \otimes 1)(x_f) = f(\varrho(u))$ for u in A*, by (16.20) and the definition of π_S. That is -

(16.21) $$\pi_S(T)(x_f) = f\varrho = \underset{\sim}{\varrho}(T)(f) \text{ in } \underline{A}(A^*,T)$$

We are now ready to define the map $\theta\colon \underset{\sim}{H} \to V_S$ by the condition -

$$\theta(T)(f) = x_f \quad (T \text{ in } |\underline{A}|,\ f \text{ in } \underline{A}(H,T))$$

The remarks above ensure that $\theta(T)$ is well-defined, and a routine computation establishes that it is natural in T. Moreover, (16.19) guarantees that $\theta(T)$ is a homomorphism of abelian groups, and thus θ is a map of abelian sheaves. We then obtain from (16.21) that the right-most square of (16.18) commutes. A final easy calculation, using the fact noted above (and in Theorem 15.10) that $\psi(z) = \varepsilon_H(z)t = \varepsilon_{S^*}(z)t$ for z in $H_1 = S^*$, yields that the left-most square likewise commutes. Therefore (16.18) is a commutative diagram, and since both rows are extensions of sheaves it follows that θ is an isomorphism. Comparison of (16.17) and (16.18) then shows that $j_A(cl(S)) = cl(\xi)$ in $\mathrm{Ext}^1_{\underline{S}}(\underset{\sim}{A^*},U)$, whence j_A is onto and the proof of Theorem 16.14 is complete.

17. Special Cases of the Theorem.

Our purpose here is to set forth some alternate forms and special cases of the isomorphism of Theorem 16.14, the first of which is the following.

Theorem 17.1. If R is a commutative local ring, then every abelian presheaf on $\underline{A}^{op}$ is a sheaf (we use, of course, the Grothendieck topology - as well as the notation and terminology - of Sections 15 and 16.) Thus, for $\underline{A}$ in $|\underline{H}_0|$, the isomorphism of Theorem 16.14 becomes -

$$j_A: X(A) \overset{\sim}{\to} \mathrm{Ext}^1_{\underline{P}}(\underset{\sim}{A^*},U)$$

with $\underline{P} = \underline{P}(\underline{A}^{op})$ the category of abelian presheaves on $\underline{A}^{op}$.

Proof. Since the elements $x_1,\dots,x_r$ of (15.1) lie outside the Jacobson radical of R, it follows that each is in U(R), since R is local. Therefore each covering of R is simply a direct product of finitely many copies of R. Since a covering of an arbitrary object S of $\underline{A}$ has the form $S \to S' = S \otimes T$, with $s \to s \otimes 1$ and T a covering of R, we obtain an S-algebra homomorphism $f: S' \to S$ (the S-algebra structure on S' arising from the first factor). If F is an abelian presheaf on $\underline{A}^{op}$, application of [13, (3.2), p. 47] (or an easy direct argument) then yields that the natural map -

$$F(S) \to \check{H}^0(S'/S,F) = \mathrm{Ker}\{F(d^0) - F(d^1) : F(S') \to F(S' \otimes_S S')\}$$

is an isomorphism, with $\check{H}^*(S'/S,F)$ the Cech or Amitsur cohomology, with coefficients in F, of the covering $S \to S'$. The theorem follows.

We turn our attention now to Hopf algebras and Galois objects over a field k, beginning with an extremely useful notion of duality which we outline without proof. For further details we refer the reader to the monograph of Heyneman and Sweedler [24].

If V is a k-space, then the natural map $V^* \otimes V^* \to (V \otimes V)^*$ is one-to-one, and we shall identify $V^* \otimes V^*$ with its image under this map. If C is a k-coalgebra, then C* is easily seen to be a k-algebra, with unit map $\varepsilon_C^*: k \to C^*$ and multiplication defined to be the composite -

$$C^* \otimes C^* \to (C \otimes C)^* \xrightarrow{\Delta_C^*} C^*$$

If C is cocommutative then C* is commutative. These observations give rise to a functor $(\ )^*: \underline{C} \to \underline{A}^{op}$.

In the remainder of this section we shall assume that all algebras, coalgebras, etc., are over a field k.

Proposition 17.2. The functor $(\ )^*: \underline{C} \to \underline{A}^{op}$ possesses a right adjoint $(\ )^D: \underline{A}^{op} \to \underline{C}$. If A is in $|\underline{A}|$, then $A^D = (\mu_A^*)^{-1}(A^* \otimes A^*) \subseteq A^*$ (with $\mu_A: A \otimes A \to A$ the multiplication map) and $\Delta_{A^D}: A^D \to A^D \otimes A^D$ is the unique map rendering the diagram below commutative -

$$\begin{array}{ccc} A^D & \xrightarrow{\Delta_{A^D}} & A^D \otimes A^D \\ \downarrow & & \downarrow \\ & & A^* \otimes A^* \\ & & \downarrow \\ A^* & \xrightarrow{\mu_A^*} & (A \otimes A)^* \end{array}$$

the unlabeled arrows denoting the inclusion maps. The augmentation of A^D is the restriction to A^D of $\eta_A^*: A^* \to k$, with $\eta_A: k \to A$ the unit map of A. If $\dim_k(A) < \infty$, then $A^D = A^*$ and $(A^D)^* = A^{**} = A$.

Remark 17.3. To say that $(\ )^D$ is right adjoint to $(\ )^*$ is simply to say that there exist set isomorphisms -

$$\underline{C}(C, A^D) \approx \underline{A}(A, C^*) \qquad (A \text{ in } |\underline{A}|,\ C \text{ in } |\underline{C}|)$$

which are natural in both variables. These are defined in the following way. If $f: A \to C^*$ in $\underline{A}$, then the corresponding $\underline{C}$-map $\bar{f}: C \to A^D$ is the unique map rendering the diagram below commutative -

$$\begin{array}{ccc} C & \xrightarrow{\bar{f}} & A^D \\ \downarrow & & \downarrow \\ C^{**} & \xrightarrow{f^*} & A^* \end{array}$$

the unlabeled arrows denoting inclusion maps. On the other hand, if $g: C \to A^D$ in $\underline{C}$, then the corresponding $\underline{A}$-map $\bar{g}: C^* \to A$ is the composite -

$$A \to A^{**} \xrightarrow{i_A^*} (A^D)^* \xrightarrow{g^*} C^*$$

where $i_A: A^D \to A^*$ and the unlabeled map are inclusions.

Now, if H is in $|\underline{H}|$, the preceding remarks show that $\underline{C}(-,H^D) \approx \underline{A}(H,(-)^*)$ is a functor from $\underline{C}^{op}$ to Ab, in which case $\underline{H}^D$ is an abelian group in the category $\underline{C}$. In virtue of the identities -

$$(17.4) \qquad \mathrm{Ab}(\underline{C}) \approx \underline{H} \approx \mathrm{Coab}(\underline{A})$$

(noted at the beginning of Section 15), we then see that H^D may be viewed as an object of $\underline{H}$. H^D will be called the Hopf dual of H. It is easy to see that the multiplication map of H is the composite -

$$(17.5) \qquad H^D \otimes H^D \to (H \otimes H)^D \xrightarrow{(\Delta_H)^D} H^D$$

the unlabeled arrow denoting the inclusion (actually an isomorphism). The augmentation and antipode of H^D are simply $\eta_H{}^D$ and $\lambda_H{}^D$, respectively, with η_H: $k \to H$ and λ_H:$H \to H$ the unit map and antipode of H, respectively. These observations yield easily the following corollary to Proposition 17.2.

Corollary 17.6. The functor $(\ )^D$: $\underline{A}^{op} \to \underline{C}$ induces a functor $(\ )^D$: $\underline{H}^{op} \to \underline{H}$. If H is in $|\underline{H}|$, then the isomorphisms of Remark 17.3 give an equivalence of functors -

$$\underline{C}(-,H^D) \approx \underline{A}(H,(-)^*) : \underline{C}^{op} \to \mathrm{Ab}$$

Finally, if H is in $|\underline{H}_0|$ then $H^D = H^*$.

It will be useful to introduce the full subcategories $\underline{A}_0$ and $\underline{C}_0$ of $\underline{A}$ and $\underline{C}$, respectively, the objects of which are all commutative k-algebras (cocommutative k-coalgebras) of finite k-dimension. We shall denote by i: $\underline{C}_0 \to \underline{C}$ the inclusion functor. The last statement of Proposition 17.2 makes clear the fact that the functor $j = (-)^*$: $\underline{C}_0 \to \underline{A}^{op}$ induces an isomorphism of $\underline{C}_0$ with $\underline{A}_0^{op}$. In particular, j is full and faithful.

Until further notice we shall write $\underline{P} = \underline{P}(\underline{A}^{op})$ and $\underline{P}' = \underline{P}(\underline{C})$, the categories of abelian presheaves on $\underline{A}^{op}$ and $\underline{C}$, respectively.

Proposition 17.7. Let A,B be in $|\underline{H}_0|$ and $|\underline{H}|$, respectively. Then there exists an isomorphism -

$$\mathrm{Ext}^1_{\underline{P}}(\underline{A}(A,-),\underline{A}(B,-)) \approx \mathrm{Ext}^1_{\underline{P}'}(\underline{C}(-,A^D),\underline{C}(-,B^D))$$

natural in A and B, which may be described as follows. If $\mathrm{cl}(\xi)$ is in $\mathrm{Ext}^1_{\underline{P}}(\underline{A}(A,-),\underline{A}(B,-)$ with -

$$\xi: 0 \to \underline{A}(B,-) \to E \to \underline{A}(A,-) \to 0$$

n extension of presheaves on $\underline{A}^{op}$, then by Theorem 14.2, there exists a sequence -

$$A \xrightarrow{\varrho} H \xrightarrow{\varphi} B$$

n $\underline{H}$ and a $\underline{P}$-isomorphism $\zeta: E \overset{\approx}{\to} \underline{A}(H,-)$ rendering the diagram below commutative -

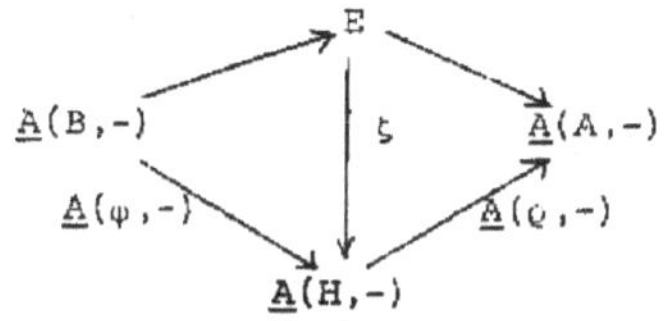

hen

$$\xi^D: 0 \to \underline{C}(-,B^D) \xrightarrow{\underline{C}(-,\varphi^D)} \underline{C}(-,H^D) \xrightarrow{\underline{C}(-,\varrho^D)} \underline{C}(-,A^D) \to 0$$

s an extension of abelian presheaves on $\underline{C}$, and $cl(\xi^D)$ is the element of $xt^1_{\underline{P}'}(\underline{C}(-,A^D),\underline{C}(-,B^D))$ corresponding to $cl(\xi)$.

Proof. We have easily from Corollary 17.6 that the following diagram of categories nd functors commutes up to an equivalence σ -

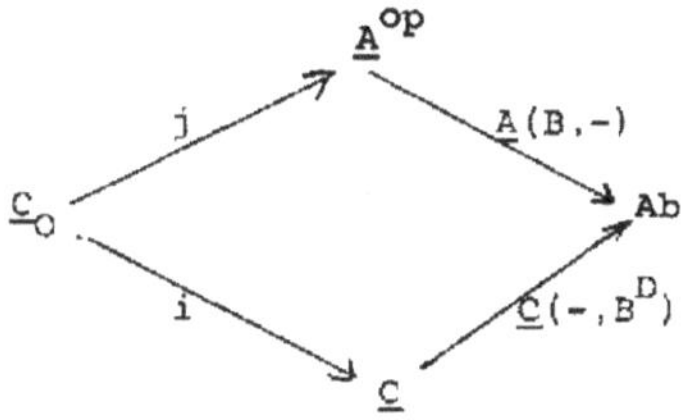

ince $\dim_k(A) < \infty$, Corollary 17.6 implies that $A^D = A^*$ in $\underline{H}$, whence $j(A^D) = A^{**} = A$ in . Therefore, since i and j are full and faithful (and clearly preserve products), we ay apply Corollary 14.11 to obtain a natural isomorphism

$$\kappa: \operatorname{Ext}^1_{\underline{P}}(\underline{A}(A,-),\underline{A}(B,-)) \xrightarrow{\approx} \operatorname{Ext}^1_{\underline{P}'}(\underline{C}(-,A^D),\underline{C}(-,B^D))$$

endering the diagram below commutative -

$$\begin{array}{ccc}
\mathrm{Ext}^1_{\underline{P}}(A(A,-),\underline{A}(B,-)) & \xrightarrow{\kappa} & \mathrm{Ext}^1_{\underline{P}'}(\underline{C}(-,A^D),\underline{C}(-,B^D)) \\
\downarrow j^* & & \downarrow i^* \\
\mathrm{Ext}^1_{\underline{P}'_0}(\underline{C}_0(-,A^D),\underline{A}(B,j(-))) & \xrightarrow{\mathrm{Ext}^1_{\underline{P}'_0}(\underline{C}_0(-,A^D),\,\sigma)} & \mathrm{Ext}^1_{\underline{P}'_0}(\underline{C}_0(-,A^D),\underline{C}(i(-),B^D))
\end{array}$$

with $\underline{P}'_0 = \underline{P}(\underline{C}_0)$. In order to verify the description of the isomorphism κ given in the statement of the theorem, one must use the information regarding σ provided in Proposition 17.2 to trace through the various maps involved; we omit the routine details.

We shall write $W = (kZ)^D$, an object of $\underline{H}$. The following fact is an easy consequenc of the definition of $(\ )^D$, as set forth in Proposition 17.2.

(17.8) $W = (kZ)^D$ is the k-algebra of all functions $w\colon Z \to k$ with the following proper There exist functions $u_1,\dots,u_r$; $v_1,\dots,v_r$ from Z to k (depending on w) such that -

$$w(m+n) = \sum_{i=1}^{r} u_i(m)\,v_i(n)$$

for all m,n in Z. Moreover, u_i and v_i can be chosen to lie in W for all $i \leq r$, in which case -

$$\Delta_W(w) = \sum_{i=1}^{r} u_i \otimes v_i$$

Putting together Theorem 17.1 and Proposition 17.7 we obtain -

<u>Theorem 17.9.</u> Given A in $|\underline{H}_0|$, there exists an isomorphism -

$$X(A) \approx \mathrm{Ext}^1_{\underline{P}(\underline{C})}(\underline{C}(-,A),\underline{C}(-,W))$$

where X(A) is, as in Section 16, the abelian group of Galois A-objects in the category $\underline{A}$. The isomorphism is natural in A.

We shall use this result to make some computations. Let A in $|\underline{H}|$ be defined as follows:

(17.10) $\quad A = k[x]$ as a k-algebra (x an indeterminate)

$$\Delta_A(x) = x \otimes 1 + 1 \otimes x$$

$$\varepsilon_A(x) = 0$$

$$\lambda_A(x) = -x$$

It is well-known, and trivially verified, that $\underline{A}(A,-) = (-)^+$, where B^+ denotes the additive group of the k-algebra B. The remark below is likewise easily checked.

Remark 17.11. If A is as in (17.10), then $\underline{H}(A,W) \approx k^+$. If a is in k^+, then the corresponding map $h_a : A \to W$ in $\underline{H}(A,W)$ is defined by the formula -

$$h_a(x)(n) = na \quad (n \text{ in } Z)$$

which makes sense in virtue of (17.8).

Now let k be a field of characteristic $p \neq 0$, and J be a (multiplicatively written) cyclic group of order p with generator σ. Let $e_0, e_1, \ldots, e_{p-1}$ be a k-base of $Jk = (kJ)^*$ dual to the base $1, \sigma, \sigma^2, \ldots, \sigma^{p-1}$ of kJ, and set $z = e_1 + 2e_2 + \ldots + (p-1)e_{p-1}$ in Jk. The lemma below is then an easy consequence of routine computations, which we omit.

Lemma 17.12. There exists a sequence in $\underline{H}$ -

$$A \xrightarrow{\wp} A \xrightarrow{\varphi} Jk$$

with A as in (17.10), such that $\wp(x) = x^p - x$ and $\varphi(z) = x$. $\wp$ is one-to-one, and $\varphi^{-1}(0)$ is the ideal of A generated by $\wp(x)$. Moreover, there exists a unique $\underline{C}$-map $\alpha : Jk \to A$ such that $\alpha(z) = x$, whence $\varphi\alpha = 1_{Jk}$.

It will be useful to analyze further the sequence of Lemma 17.12, employing the fact, proved by Grothendieck, that $\underline{H}$ is an abelian category (see [39, §3.3, p. 9] for a discussion, although not a proof of this theorem). We show that the sequence is, in some sense, a projective resolution of Jk.

Proposition 17.13. (a) The sequence -

$$0 \to \underline{C}(-,A) \xrightarrow{\underline{C}(-,\wp)} \underline{C}(-,A) \xrightarrow{\underline{C}(-,\varphi)} \underline{C}(-,Jk) \to 0$$

with A, etc., as in Lemma 17.12, is a s.e.s. in $\underline{P}(\underline{C})$.

(b) $\mathrm{Ext}^1_{\underline{P}(\underline{C})}(\underline{C}(-,A), \underline{C}(-,B)) = 0$ for any B in $|\underline{H}|$.

Proof. Lemma 17.12, together with the discussion of $\underline{H}$ to be found in [39, §3.2 and 3.3, p. 9], yield easily that $\wp$ is a monomorphism in $\underline{H}$ and φ is its cokernel. Thus, since $\underline{H}$ is an abelian category, $\wp$ is the kernel of φ, whence by [39, §3.2, p. 9] the diagram -

$$A \xrightarrow{\wp} A \overset{\varphi}{\underset{\varepsilon}{\rightrightarrows}} Jk$$

is an equalizer diagram in $\underline{C}$, with ε the composition -

$$A \xrightarrow{\varepsilon_A} k \to Jk$$

Now consider the $\underline{C}$-map $\theta = 1_A - \alpha\varphi\colon A \to A$ (the minus sign making sense in virtue of the fact that $\underline{C}(-,A)$ is a functor from $\underline{C}$ to Ab.) Since $\varphi\alpha = 1_{Jk}$ and ε is the zero element of the abelian group $\underline{C}(A,Jk)$, we have that $\varphi\theta = \varepsilon = \varepsilon\theta$. Thus, by the universal property of equalizers, there is a unique $\underline{C}$-map $\nu\colon A \to A$ such that $\wp\nu = \theta$, or $\wp\nu + \alpha\varphi = 1_A$. Furthermore, $\wp\nu\wp = (1_A - \alpha\varphi)\wp = \wp - \alpha\varepsilon = \wp$, whence $\nu\wp = 1_A$ because $\wp$ is a monomorphism in $\underline{H}$. (a) then follows easily from Lemma 14.1(c).

Turning now to (b), Theorem 14.2 guarantees that we may assume without loss of generality that every element of $\mathrm{Ext}^1_{\underline{P}(\underline{C})}(\underline{C}(-,A),\underline{C}(-,B))$ is represented by an extension of the form -

$$\xi\colon 0 \to \underline{C}(-,B) \xrightarrow{\underline{C}(-,\psi)} \underline{C}(-,H) \xrightarrow{\underline{C}(-,\varrho)} \underline{C}(-,A) \to 0$$

with $B \xrightarrow{\psi} H \xrightarrow{\varrho} A$ a sequence in $\underline{H}$. Then, by Lemma 14.1(c) and the Yoneda Lemma, there exists a $\underline{C}$-map $\tau\colon A \to H$ such that $\varrho\tau = 1_A$. If $\tau(x) = y$, then (17.10) implies that $\Delta_H(y) = y \otimes 1 + 1 \otimes y$. Therefore, if $\bar{\tau}\colon A = k[x] \to H$ is the k-algebra homomorphism such that $\bar{\tau}(x) = y$, we see that $\bar{\tau}$ is an $\underline{H}$-map and $\varrho\bar{\tau} = 1_A$, because ϱ is likewise an $\underline{H}$-map. This means that the s.e.s. ξ splits; i.e., $\mathrm{cl}(\xi) = 0$ and the proof of the Proposition is complete.

The exact sequence of Proposition 17.13(a), together with Remark 17.11, give rise to the diagram -

$$\begin{array}{ccccccc}
\underline{P}'(\underset{\sim}{A},\underset{\sim}{W}) & \xrightarrow{\underset{\sim}{\wp}^*} & \underline{P}'(\underset{\sim}{A},\underset{\sim}{W}) & \xrightarrow{\delta} & \mathrm{Ext}^1_{\underline{P}'}(\underset{\sim}{Jk},\underset{\sim}{W}) & \xrightarrow{\underset{\sim}{\wp}^*} & \mathrm{Ext}^1_{\underline{P}'}(\underset{\sim}{A},\underset{\sim}{W}) \\
\| & & \| & & & & \| \\
\underline{H}(A,W) & \xrightarrow{\underline{H}(\wp,W)} & \underline{H}(A,W) & & & & 0 \\
\| & & \| & & & & \\
k^+ & \longrightarrow & k^+ & & & &
\end{array}$$

where $\underline{P}' = \underline{P}(\underline{C})$ and $\underset{\sim}{A} = \underline{C}(-,A)$, etc. The top row in the diagram is the exact cohomology sequence of the s.e.s. of (17.13a), the upper vertical isomorphisms arise from the Yoneda Lemma and Proposition 17.13(b), respectively, and the lower ones are those of Remark 17.11. It is easily verified that the induced map $k^+ \to k^+$ satisfies the formula -

$$a \to a^p - a \qquad (a \text{ in } k)$$

whence we obtain, via Theorem 17.9 and Example 12.6, the following fact.

Corollary 17.14. If k is a field of characteristic $p \neq 0$ and J is a cyclic group of order p, then -

$$\mathrm{Hom}_c(\Pi,J) \approx X(Jk) \approx k^+/\{a^p-a\}$$

with Π the Galois group of a separable closure of k, and the left-hand side denoting continuous homomorphisms from the compact group Π to the discrete group J.

Corollary 17.14 is the well-known description of normal separable extensions of degree p of a field of characteristic p. Our proof is substantially more complicated than the standard one using Galois cohomology (see, e.g., [37, Chapitre X, §3, (a), p. 163].) But our purpose here was to provide some insight into the relation between this special result and the general isomorphism of Theorem 17.1.

We turn finally to a generalization of the Kummer isomorphism [37, Chapitre X, §3, (b)]. In the remainder of this section k will be a field of characteristic prime to a fixed positive integer n. In order to interpret our results in this case, it will be desirable to compare the category $\underline{A}$ of commutative k-algebras to a certain category of sets which we now introduce. Let K be the splitting field over k of the equation $x^n - 1 = 0$, and Γ be the Galois group of the normal separable extension $K|k$. We shall denote by U_n the set of all n'th roots of 1 in K, a cyclic subgroup of U(K) of order n which is, moreover, a Γ-module in the obvious way.

Now let $(\mathrm{Sets})^\Gamma$ be the category of Γ-sets (i.e., sets equipped with Γ-action), and $(\mathrm{Sets})_0^\Gamma$ be the full subcategory of which the objects are the finite Γ-sets. We introduce the following convenient notation:

$$(X,Y) = \mathrm{Sets}(X,Y) \quad \text{for } X,Y \text{ in } |\mathrm{Sets}|$$
$$(X,Y)^\Gamma = (\mathrm{Sets})^\Gamma(X,Y) \quad \text{for } X,Y \text{ in } |(\mathrm{Sets})^\Gamma|$$

"Sets" is, of course, the category of sets. Observe now the existence of the product-preserving functors -

$$i: (\mathrm{Sets})_0^\Gamma \to (\mathrm{Sets})^\Gamma$$
$$j: (\mathrm{Sets})_0^\Gamma \to \underline{A}^{op}$$

with i the inclusion functor and $j = (-,K)^\Gamma$, with k-algebra structure arising from the pointwise operations. i is, of course, full and faithful. In order to analyze j, we introduce the full subcategory $\underline{A}_s$ of $\underline{A}$, the objects of which are the commutative k-algebras which are finite direct products of subfields of K. If X is in $|(\mathrm{Sets})_0^{\Gamma}|$, an easy computation yields that $(X,K)^{\Gamma}$ is in $|\underline{A}_s|$, whence j induces a functor j_1: $(\mathrm{Sets})_0^{\Gamma} \to \underline{A}_s^{op}$. If A is in $|\underline{A}_s|$, then $\underline{A}(A,K)$ is a finite Γ-set, the Γ-structure arising from that of K. Thus we obtain a functor $j_2 = \underline{A}(-,K): \underline{A}_s^{op} \to (\mathrm{Sets})_0^{\Gamma}$. A routine exercise in Galois theory then yields the following facts.

Lemma 17.15. j_1 and j_2 are category isomorphisms, and each is naturally equivalent to the inverse of the other. Thus j is full and faithful.

Notation and Remarks 17.16. Set $\underline{M}^{\Gamma}$ and $\underline{M}_0^{\Gamma}$ denote the categories of Γ-modules and finite Γ-modules, respectively. Then $\underline{M}^{\Gamma} = \mathrm{Ab}((\mathrm{Sets})^{\Gamma})$ and $\underline{M}_0^{\Gamma} = \mathrm{Ab}((\mathrm{Sets})_0^{\Gamma})$, whence Lemma 17.15 yields category isomorphisms -

$$j_1: \underline{M}_0^{\Gamma} \xrightarrow{\approx} \mathrm{Ab}(\underline{A}_s^{op}) = \mathrm{Coab}(\underline{A}_s)^{op}$$

$$j_2: \mathrm{Coab}(\underline{A}_s)^{op} = \mathrm{Ab}(\underline{A}_s^{op}) \xrightarrow{\approx} \underline{M}_0^{\Gamma}$$

each being naturally equivalent to the inverse ot the other. Of course, $\mathrm{Coab}(\underline{A}_s) = \underline{H}_s$ is simply the full subcategory of $\underline{H}$ of which the objects are all finite commutative, cocommutative Hopf algebras which, as algebras, are direct products of subfields of K. In particular, if M is a finite Γ-module, then $(M,K)^{\Gamma}$ is a Hopf k-algebra with these properties, with coalgebra operations arising in an obvious way from the addition, etc., in M.

Lemma 17.17. Let J be a finite abelian group of exponent n. Then $\mathrm{Hom}_Z(J,U_n)$ is a finite Γ-module (the Γ-structure arising from that of U_n), and there exists an $\underline{H}$-isomorphism α: $kJ \xrightarrow{\approx} (\mathrm{Hom}_Z(J,U_n),K)^{\Gamma}$ such that

$$\alpha\Big(\sum_{\sigma \text{ in } J} c_\sigma \sigma\Big)(f) = \sum_{\sigma \text{ in } J} c_\sigma f(\sigma)$$

for c_σ in k, f in $\mathrm{Hom}_Z(J,U_n)$. α is natural in J.

Proof. Since J has exponent n, and K has characteristic prime to n and contains a all n'th roots of 1, it is well known and trivially verified that every irreducible representation of J in K is one-dimensional, whence kJ is in $|\underline{H}_s|$. The observation that a k-algebra homomorphism $kJ \to K$ is the linear extension of a unique group homomorphism $J \to U_n$ then yields a Γ-module isomorphism -

$$\mathrm{Hom}_Z(J,U_n) \xrightarrow{\approx} \underline{A}(kJ,K)$$

and application of j_1 and (17.16) then yields the desired $\underline{H}$-isomorphism α. The formula for α given is easily checked.

<u>Theorem 17.18.</u> Let k, K, Γ, U_n be as in the preceding discussion, J be a finite abelian group of exponent n, $\underline{P} = \underline{P}(\underline{A}^{op})$, and $\underset{\sim}{H} = \underline{A}(H,-)$ for H in $|\underline{H}|$. Then there exists an isomorphism -

$$\mathrm{Ext}^1_{\underline{P}}(\underset{\sim}{kJ},U) \approx \mathrm{Ext}^1_{Z\Gamma}(\mathrm{Hom}_Z(J,U_n),U(K))$$

which is natural in J.

Proof. Observe first that the diagram below of categories and functors commutes -

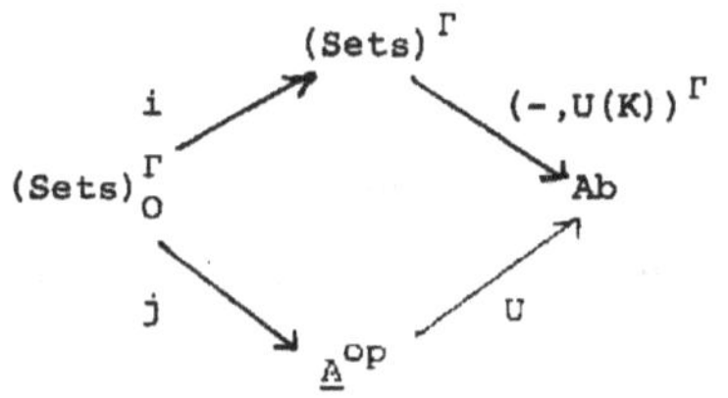

We have seen in Lemma 17.15 that j is full and faithful, as is i. Moreover, it is an easy exercise to show that both i and j preserve products. We are thus in the situation of Corollary 14.11, which we apply (with $M = \mathrm{Hom}_Z(J,U_n)$ in $|\underline{M}^{\Gamma}_0|$ playing the role of A) to obtain a unique isomorphism -

$$\kappa: \mathrm{Ext}^1_{\underline{P}'}((-,M)^{\Gamma},(-,U(K))^{\Gamma}) \xrightarrow{\approx} \mathrm{Ext}^1_{\underline{P}}(\underset{\sim}{j(M)},U)$$

rendering the diagram below commutative -

$$\begin{array}{ccc}
\mathrm{Ext}^1_{\underline{P}'}((-,M)^{\Gamma},(-,U(K))^{\Gamma}) & \xrightarrow{\kappa} & \mathrm{Ext}^1_{\underline{P}}(\underset{\sim}{j(M)},U) \\
\big\downarrow i^* & & \big\downarrow j^* \\
\mathrm{Ext}^1_{\underline{P}'_0}((-,M)^{\Gamma},(-,U(K))^{\Gamma}) & = & \mathrm{Ext}^1_{\underline{P}'_0}((-,M)^{\Gamma},Uj)
\end{array}$$

where $\underline{P}' = \underline{P}((\mathrm{Sets})^{\Gamma})$ and $\underline{P}'_0 = \underline{P}((\mathrm{Sets})^{\Gamma}_0)$. We also have the isomorphism -

$$\mathrm{Ext}^1_{\underline{P}}(\underset{\sim}{\alpha},U): \mathrm{Ext}^1_{\underline{P}}(\underset{\sim}{kJ},U) \xrightarrow{\approx} \mathrm{Ext}^1_{\underline{P}}(\underset{\sim}{j(M)},U)$$

with $\alpha: kJ \overset{\approx}{\to} j(M)$ the natural $\underline{H}$-isomorphism of Lemma 17.17. Finally, we note that, by

Hilbert's Theorem 90 [37, Proposition 2, p. 158], $H^1(\Gamma',U(K)) = \mathrm{Ext}^1_{Z\Gamma'}(Z,U(K)) = 0$ for every subgroup Γ' of Γ, whence Example 14.12 yields an isomorphism -

$$\theta\colon \mathrm{Ext}^1_{\underline{P}'}((-,M)^{\Gamma},(-,U(K))^{\Gamma}) \xrightarrow{\approx} \mathrm{Ext}^1_{Z\Gamma}(M,U(K))$$

Putting these three maps together, we obtain the desired isomorphism.

Corollary 17.19. Let k be a field of characteristic prime to a natural number n, K be the splitting field over k of the equation $x^n - 1 = 0$, and U_n be the multiplicative group of n'th roots of 1 in K. Let Π be the Galois group of a separable closure of k, Γ be the Galois group of the extension $K|k$, and J be a finite abelian group of exponent n. Then there exist isomorphisms -

$$\mathrm{Hom}_c(\Pi,J) \approx X(Jk) \approx \mathrm{Ext}^1_{Z\Gamma}(\mathrm{Hom}_Z(J,U_n),U(K))$$

natural in J, the left-hand side denoting continuous homomorphisms from the profinite group Π to the discrete group J. The left-most isomorphism is that of Example 12.6, whereas the right-most one arises from Theorems 17.1 and 17.18.

For the special case in which K = k and $J = U_n$, we recapture easily the Kummer isomorphism of [37, Chapitre X, §3, (b)].

Remarks

See also J. Giraud's torseurs in a topos [19].

Recently Mr. K. Newman, of Cornell University, has generalized the author's proof for the special case $H = RG$ to obtain the following result: Let H be a group in $\underline{C}$ with the property that the R-submodule of integrals [26] of H* consists of cocommutative elements. Then a Galois H-object in $\underline{C}$ is a finitely generated projective R-module.

See [9, Section 1, Theorem 1].

In order to illustrate the theory developed here, we shall outline an example of an integral extension of the ring Z of rational integers which is a Galois A-object for a suitable finite Hopf Z-algebra A.

Let $K = Q(\zeta)$, where Q is the field of rational numbers and ζ is a primitive q'th root of 1, where $q = 2^{n+1}$ for some $n > 0$. Since the powers of ζ are precisely the roots of the irreducible cyclotomoc polynomial $\Phi_q(t) = t^{2^n} + 1$, K is a normal extension of Q of degree 2^n. One may then apply standard techniques of algebraic number theory to obtain that the integral closure S of Z in K is a free abelian group with basis $1, \zeta, \zeta^2, \dots, \zeta^{2^n - 1}$, hence is a faithfully flat Z-module.

Now let $J = Z/2^n Z$, an abelian group which we shall write multiplicatively with generator τ. It is then trivially verified that there exists a unique Z-algebra homomorphism $\alpha_S : S \to S \otimes ZJ$ such that $\alpha_S(\zeta) = \zeta \otimes \tau$; furthermore, this map gives to S the structure of a ZJ-object. (Here ZJ is the integral group ring of J, a finite Hopf Z-algebra as explained on p.59). An easy computation shows that the map $\gamma_S : S \otimes S \to S \otimes ZJ$ is onto, hence an isomorphism by a routine rank argument. We may then conclude that S is a Galois ZJ-object. (Of course, S is a special case of Example 4.16; the corresponding pair (I, β) satisfies the conditions:
$I = Z$, $\beta(x_1 \otimes \dots \otimes x_{2^n}) = - x_1 \dots x_{2^n}$.)

It is not difficult to show that a subring T of S corresponds, via Theorem 7.6, to a finite Hopf subalgebra of $(ZJ)^*$ if and only if $T = Z[\zeta^{2^k}]$ for some $k \leq n$.

5. The argument just given is inadequate. The following is better: In the first paragraph after the proof of Proposition 1 of [26], there is exhibited a Hopf module isomorphism $A^* \approx A \otimes P$, with A and A* playing the roles of H and H*, respectively. In particular, the isomorphism is a left A-comodule isomorphism, where the left A-comodule structure of A* arises from the natural right A*-module structure of A* via an adjointness relation analogous to that of (7.2), and the left A-comodule structure of A is given by Δ_A. A moment's reflection on (7.2), and the remark which follows it, shows that the above isomorphism $A^* \approx A \otimes P$ is a right A*-module isomorphism, where A* has the natural right A*-module structure, and A has the right A*-module structure induced, via duality, by the natural left A*-module structure of A*. Tensoring with $I' = P^{-1}$ then gives $A \approx A^* \otimes I'$ as right A*-modules.

6. The computation in the proof of part (a) of this theorem can be avoided by a scrutiny of the functors represented by T, A, B. Let $\underline{A}$ be the category of commutative R-algebras, and set $F = \underline{A}(T,-)$, $G = \underline{A}(A,-)$, $H = \underline{A}(B,-)$. F is a functor from $\underline{A}^{op}$ to Sets, and G and H are functors from $\underline{A}^{op}$ to the category of groups. The surjection $f: A \to B$ gives rise to an injection $H(C) \to G(C)$ for any C in $|\underline{A}|$, and we identify H(C) with its image in G(C). The equalizer diagram -

$$T \longrightarrow A \overset{i}{\underset{\alpha'_A}{\rightrightarrows}} A \otimes B$$

of Proposition 10.1 yields a coequalizer diagram of set-valued functors -

$$G \times H \rightrightarrows G \overset{P}{\longrightarrow} F$$

where, for any C as above, the induced maps $(G \times H)(C) \approx G(C) \times H(C) \to G(C)$ are defined by $(\sigma,\tau) \to \sigma$ and $(\sigma,\tau) \to \sigma\tau$ for σ in G(C), τ in H(C). This coequalizer diagram in turn gives rise to a natural transformation $\gamma: G \times H \to G \times_F G$ of set-valued functors. Now, the functor $G \times_F G$ is represented by $A \otimes_T A$, from whence one sees easily that γ is induced by the homomorphism -

$$\gamma'_A: A \otimes_T A \to A \otimes_T (T \otimes B) \approx A \otimes B$$

of Definition 7.3, which is an isomorphism by Corollary 10.4. Thus γ is an equivalence of functors. That is, the mapping -

$$\gamma(C): G(C) \times H(C) \to G(C) \times_{F(C)} G(C)$$

is an isomorphism for any C in |A|. An easy computation then shows that the diagram below is an equalizer diagram of sets -

$$H(C) \to G(C) \underset{p_1(C)}{\overset{p(C)}{\rightrightarrows}} F(C)$$

where $p_1(C)(\sigma) = p(C)(1)$ for σ in $G(C)$. Thus -

$$H \to G \underset{p_1}{\overset{p}{\rightrightarrows}} F$$

is an equalizer diagram of functors. But it is easy to see that this diagram is represented by the diagram -

$$T \rightrightarrows A \xrightarrow{f} B$$

in $\underline{A}$, the maps $T \to A$ being the inclusion and $x \to \epsilon_A(x)$, respectively. This diagram must therefore be a coequalizer diagram in $\underline{A}$, from which condition (a) of Theorem 10.6 follows immedately.

7. A more detailed argument is needed here. If $aJ_{A^*} = 0$ for some a in A, then $W J_{A^*} = 0$, where W is the image in A* of $M^{-1} \otimes a$ under the right A*-module isomorphism $M^{-1} \otimes A \xrightarrow{\approx} A^*$. If w is in W, then the definition of J_{A^*} guarantees that $\epsilon_{A^*}(w) J_{A^*} = 0$. We may then apply, for example, Corollary 9.7 (with A playing the role of S) to obtain that $\epsilon_{A^*}(w) = 0$, whence w is in I_{A^*}. It follows that $W \subseteq I_{A^*}$, and therefore a is in the image of $M \otimes I_{A^*}$ in A under the isomorphism $M \otimes A^* \xrightarrow{\approx} A$, which is clearly in AI_{A^*}.

8. In this chapter we shall denote by X the functor $\mathbf{X}$ defined in Chapter I, Section 3.

9. After these notes were typed, the interesting work of H.Epp [46] was brought to our attention. Epp, using a different method, establishes the isomorphism exhibited above - using sheaves in the faithfully flat topology - for the special case in which A* is of multiplicative type (he does not assume, however, that A is a finitely generated R-module).

10. Of course, AZ is an RZ-object via the formula $\alpha_{AZ}(at^n) = at^n \otimes t^n$ for a in A. The properties of the isomorphism $H \simeq AZ$ (which we have viewed as an identification) can be explained as follows. In the context and notation of Theorem 14.2, $A \times B$ has the structure of an abelian group in $\underline{C}$, this structure being induced by the

isomorphism $\zeta: E \xrightarrow{\approx} \underline{C}(-, A\times B)$. Letting $\mu: (A\times B)\times(A\times B) \to A\times B$ be the multiplication map, the definition of ζ and the properties of the maps α and β of Lemma 14.1(c) yield without difficulty the commutativity of the diagram below -

$$\begin{array}{ccc} (A\times B)\times B & \xrightarrow{(A\times B)\times \nu} & (A\times B)\times (A\times B) \\ \| & & \downarrow \mu \\ A\times (B\times B) & \xrightarrow{A\times \mu_B} & A\times B \end{array}$$

where $\nu: B \to A\times B$ is as in Theorem 14.2(b). Translating to the context of Lemma 15.8, and keeping in mind the fact that the isomorphism $H \approx AZ$ arises from the map ζ above, we obtain the commutative diagram -

$$\begin{array}{ccc} H \approx A\otimes RZ & \xrightarrow{A\otimes \Delta_{RZ}} & (A\otimes RZ)\otimes RZ \\ \downarrow \Delta_H & \searrow \alpha_H & \| \\ H\otimes H & \xrightarrow{H\otimes \varphi} & H\otimes RZ \end{array}$$

whence $H \approx AZ$ as RZ-objects.

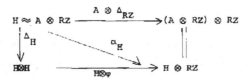

REFERENCES

[1] M. André, Méthode Simpliciale en Algèbre Homologique et Algèbre Commutative, Lecture Notes in Mathematics, Vol. 32 (ETH series), Springer-Verlag, Berlin (1967).

[2] M. Artin, Grothendieck Topologies, Harvard University mimeographed notes, Cambridge, Massachusetts (1962).

[3] M. Auslander and D. Buchsbaum, On Ramification Theory in Noetherian Rings, Amer. J. Math., Vol. 81 (1959), pp. 749-765.

[4] M. Auslander and O. Goldman, Maximal Orders, Trans. Amer. Math. Soc., Vol. 97 (1960), pp. 1-24.

[5] M. Auslander and O. Goldman, The Brauer Group of a Commutative Ring, Trans. Amer. Math. Soc., Vol. 97 (1960), pp. 367-409.

[6] R. Baer, Erweiterungen von Gruppen und ihren Isomorphismen, Math. Z., Vol. 38 (1934), pp. 375-416.

[7] M. Barr and J. Beck, Acyclic Models and Triples, Procedings of the Conference on Categorical Algebra, La Jolla (1965), Springer-Verlag, New York Inc. (1966), pp. 336-343.

[8] H. Bass, The Morita Theorems, University of Oregon mimeographed notes, Eugene, Oregon (1962).

[9] J. Beck, Triples, Algebras, and Cohomology, Ph.D. Thesis, Columbia University, New York (1965).

[10] N. Bourbaki, Algèbre Commutative, Chapters I - II, Hermann, Paris (1962).

[11] H. Cartan and S. Eilenberg, Homological Algebra, Princeton University Press, Princeton, New Jersey (1956).

[12] S.U. Chase, D.K. Harrison, and Alex Rosenberg, Galois Theory and Galois Cohomology of Commutative Rings, Memoirs Amer. Math. Soc., Vol. 52 (1965), pp. 15-33.

[13] S.U. Chase and Alex Rosenberg, Amitsur Cohomology and the Brauer Group, Memoirs Amer. Math. Soc., Vol. 52 (1965), pp. 34-79.

[14] S.U. Chase and Alex Rosenberg, A Theorem of Harrison, Kummer Theory, and Galois

Algebras, Nagoya Math. J., Vol. 27 (1966), pp. 663-685.

[15] S.U. Chase, Abelian Extensions and a Cohomology Theory of Harrison, Proceedings of the Conference on Categorical Algebra, La Jolla (1965), Springer-Verlag, New York Inc. (1966), pp. 375-403.

[16] S. Eilenberg and J.C. Moore, Adjoint Functors and Triples, Ill. Math. J., Vol. 9 (1965), pp. 381-398.

[17] P. Freyd, Abelian Categories, Harper and Row, New York, New York (1964).

[18] P. Gabriel, Des Catégories Abeliennes, Bull. Soc. Math. France, Vol. 90 (1962), pp. 323-448.

[19] J. Giraud, Cohomologie Non-Abelienne, Columbia University Notes (1965).

[20] A. Grothendieck, Technique de Descente et Théorèmes d'Existence en Géometrie Algebrique I: Généralités- Descente par Morphismes Fidèlement Plats, Seminaire Bourbaki Vol. 12 (1959/60) Expose 190.

[21] A. Grothendieck, Fondements de la Géometrie Algebrique (extraits du Seminaire Bourbaki (1957/62)), Paris (1962).

[22] D.K. Harrison, Abelian Extensions of Arbitrary Fields, Trans. Amer. Math. Soc., Vol 106 (1963), pp. 230-235.

[22']D.K. Harrison, Abelian Extensions of Commutative Rings, Memoirs Amer. Math. Soc. 52 (1965), pp. 1-14.

[23] H. Hasse, Invariante Kennzeichnung Galoisscher Körper mit vorgegebener Galoisgruppe J. Reine Angewandte Math., Vol. 187 (1950), pp. 14-43.

[24] R. Heynemann and M.E. Sweedler, Affine Hopf Algebras, to appear.

[25] G. Hochschild, On the Cohomology Groups of an Associative Algebra, Ann. of Math., Vol 46 (1945), pp. 58-67.

[26] R.G. Larson and M.E. Sweedler, An Associative Orthogonal Bilinear Form for Hopf Algebras, to appear in Amer. J. Math.

[27] S. Mac Lane, Homology, Academic Press, New York (1963).

[27'] S. Mac Lane, Categorical Algebra, Bull. Amer. Math. Soc., Vol. 71 (1965), pp. 40-10

[28] J.W. Milnor and J.C. Moore, On the Structure of Hopf Algebras, Ann. Math., Vol. 81 (1965), pp. 211-264.

29] B. Mitchell, Theory of Categories, Academic Press, New York, New York (1965).

30] K. Morita, Duality for Modules, Science Reports Tokyo Kyoiku Daigaku sect. A, Vol. 6 (1958).

31] F. Oort, Commutative Group Schemes, Lecture Notes in Mathematics, Vol. 15, Springer-Verlag, Berlin (1966).

32] M. Orzech, A Cohomology Theory for Commutative Galois Extensions, Math. Z., Vol. 105 (1968), pp. 128-140.

33] M. Orzech, A Cohomological Description of Abelian Galois Extensions, to appear.

34] Alex Rosenberg and D. Zelinsky, Cohomology of Infinite Algebras, Trans. Amer. Math. Soc., Vol. 82 (1956), pp. 85-98.

35] S. Shatz, Cohomology of Artinian Group Schemes Over Local Fields, Ann. of Math., Vol. 79 (1964), pp. 411-449.

36] J.P. Serre, Groupes Algébrique et Corps de Classes, Hermann, Paris (1959), Act. Sci. Ind. 1264.

37] J.P. Serre, Corps Locaux, Hermann, Paris (1962), Act. Sci. Ind. 1296.

38] J.P. Serre, Cohomologie Galoisienne, Lecture Notes in Mathematics, Vol. 5, Springer-Verlag, Berlin (1965).

39] P.K. Sharma, Structure Theory of Commutative Affine Groups, Séminaire Heidelberg-Strasbourg Annee 1965/66, Expose 11.

40] U. Shukla, Cohomologie des Algèbres Associatives, Ann. Sci. Ecole Norm. Sup., Vol. 78 (1961), pp. 163-209.

41] Strasbourg Univerisity Department of Mathematics, Groupes Algebriques, Seminaire Heidelberg-Strasbourg Annee 1965/66.

42] M.E. Sweedler, Cohomology of Algebras Over Hopf Algebras, Trans. Amer. Math. Soc., Vol. 133 (1968), pp. 205-239.

43] M. E. Sweedler, The Hopf Algebra of an Algebra as Applied to Field Theory, J. of J. of Algebra, Vol. 8 (1968), pp. 262-276.

44] P. Wolf, Algebraische Theorie der Galoisschen Algebren, Math. Forschungsberichte III. Berlin: Deutscher Verlag der Wissenschaften (1956).

45] G.S. Rinehart, Satellites and Cohomology, to appear.

46] H.Epp, Group Schemes, Harrison's Theorem, and Galois Extensions, Ph. D. Thesis, Northwestern University, 1966.

Offsetdruck: Julius Beltz, Weinheim/Bergstr

Lecture Notes in Mathematics

Bisher erschienen/Already published

Vol. 1: J. Wermer, Seminar über Funktionen-Algebren. IV, 30 Seiten. 1964. DM 3,80 / 0.95

Vol. 2: A. Borel, Cohomologie des espaces localement compacts d'après J. Leray. IV, 93 pages. 1964. DM 9,– / $ 2.25

Vol. 3: J. F. Adams, Stable Homotopy Theory. 2nd. revised edition. IV, 78 pages. 1966. DM 7,80 / $ 1.95

Vol. 4: M. Arkowitz and C. R. Curjel, Groups of Homotopy Classes. 2nd. revised edition. IV, 36 pages. 1967. DM 4,80 / $ 1.20

Vol. 5: J.-P. Serre, Cohomologie Galoisienne. Troisième édition. VIII, 214 pages. 1965. DM 18,– / $ 4.50

Vol. 6: H. Hermes, Eine Termlogik mit Auswahloperator. IV, 42 Seiten. 1965. DM 5,80 / $ 1.45

Vol. 7: Ph. Tondeur, Introduction to Lie Groups and Transformation Groups. VIII, 176 pages. 1965. DM 13,50 / $ 3.40

Vol. 8: G. Fichera, Linear Elliptic Differential Systems and Eigenvalue Problems. IV, 176 pages. 1965. DM 13,50 / $ 3.40

Vol. 9: P. L. Ivănescu, Pseudo-Boolean Programming and Applications. IV, 50 pages. 1965. DM 4,80 / $ 1.20

Vol. 10: H. Lüneburg, Die Suzukigruppen und ihre Geometrien. VI, 111 Seiten. 1965. DM 8,– / $ 2.00

Vol. 11: J.-P. Serre, Algèbre Locale. Multiplicités. Rédigé par P. Gabriel. Seconde édition. VIII, 192 pages. 1965. DM 12,– / $ 3.00

Vol. 12: A. Dold, Halbexakte Homotopiefunktoren. II, 157 Seiten. 1966. DM 12,– / $ 3.00

Vol. 13: E. Thomas, Seminar on Fiber Spaces. IV, 45 pages. 1966. DM 4,80 / $ 1.20

Vol. 14: H. Werner, Vorlesung über Approximationstheorie. IV, 184 Seiten und 12 Seiten Anhang. 1966. DM 14,– / $ 3.50

Vol. 15: F. Oort, Commutative Group Schemes. VI, 133 pages. 1966. DM 9,80 / $ 2.45

Vol. 16: J. Pfanzagl and W. Pierlo, Compact Systems of Sets. IV, 48 pages. 1966. DM 5,80 / $ 1.45

Vol. 17: C. Müller, Spherical Harmonics. IV, 46 pages. 1966. DM 5,– / $ 1.25

Vol 18: H.-B. Brinkmann und D. Puppe, Kategorien und Funktoren. XII, 107 Seiten, 1966. DM 8,– / $ 2.00

Vol. 19: G. Stolzenberg, Volumes, Limits and Extensions of Analytic Varieties. IV, 45 pages. 1966. DM 5,40 / $ 1.35

Vol. 20: R. Hartshorne, Residues and Duality. VIII, 423 pages. 1966. DM 20,– / $ 5.00

Vol. 21: Seminar on Complex Multiplication. By A. Borel, S. Chowla, C. S. Herz, K. Iwasawa, J.-P. Serre. IV, 102 pages. 1966. DM 8,– / $ 2.00

Vol. 22: H. Bauer, Harmonische Räume und ihre Potentialtheorie. IV, 175 Seiten. 1966. DM 14,– / $ 3.50

Vol. 23: P. L. Ivănescu and S. Rudeanu, Pseudo-Boolean Methods for Bivalent Programming. 120 pages. 1966. DM 10,– / $ 2.50

Vol. 24: J. Lambek, Completions of Categories. IV, 69 pages. 1966. DM 6,80 / $ 1.70

Vol. 25: R. Narasimhan, Introduction to the Theory of Analytic Spaces. IV, 143 pages. 1966. DM 10,– / $ 2.50

Vol. 26: P.-A. Meyer, Processus de Markov. IV, 190 pages. 1967. DM 15,– / $ 3.75

Vol. 27: H. P. Künzi und S. T. Tan, Lineare Optimierung großer Systeme. VI, 121 Seiten. 1966. DM 12,– / $ 3.00

Vol. 28: P. E. Conner and E. E. Floyd, The Relation of Cobordism to K-Theories. VIII, 112 pages. 1966. DM 9,80 / $ 2.45

Vol. 29: K. Chandrasekharan, Einführung in die Analytische Zahlentheorie. VI, 199 Seiten. 1966. DM 16,80 / $ 4.20

Vol. 30: A. Frölicher and W. Bucher, Calculus in Vector Spaces without Norm. X, 146 pages. 1966. DM 12,– / $ 3.00

Vol. 31: Symposium on Probability Methods in Analysis. Chairman. D. A. Kappos. IV, 329 pages. 1967. DM 20,– / $ 5.00

Vol. 32: M. André, Méthode Simpliciale en Algèbre Homologique et Algèbre Commutative. IV, 122 pages. 1967. DM 12,– / $ 3.00

Vol. 33: G. I. Targonski, Seminar on Functional Operators and Equations. IV, 110 pages. 1967. DM 10,– / $ 2.50

Vol. 34: G. E. Bredon, Equivariant Cohomology Theories. VI, 64 pages. 1967. DM 6,80 / $ 1.70

Vol. 35: N. P. Bhatia and G. P. Szegö, Dynamical Systems. Stability Theory and Applications. VI, 416 pages. 1967. DM 24,– / $ 6.00

Vol. 36: A. Borel, Topics in the Homology Theory of Fibre Bundles. VI, 95 pages. 1967. DM 9,– / $ 2.25

Vol. 37: R. B. Jensen, Modelle der Mengenlehre. X, 176 Seiten. 1967. DM 14,– / $ 3.50

Vol. 38: R. Berger, R. Kiehl, E. Kunz und H.-J. Nastold, Differentialrechnung in der analytischen Geometrie IV, 134 Seiten. 1967. DM 12,– / $ 3.00

Vol. 39: Séminaire de Probabilités I. II, 189 pages. 1967. DM 14,– / $ 3.50

Vol. 40: J. Tits, Tabellen zu den einfachen Lie Gruppen und ihren Darstellungen. VI, 53 Seiten. 1967. DM 6.80 / $ 1.70

Vol. 41: A. Grothendieck, Local Cohomology. VI, 106 pages. 1967. DM 10.– / $ 2.50

Vol. 42: J. F. Berglund and K. H. Hofmann, Compact Semitopological Semigroups and Weakly Almost Periodic Functions. VI, 160 pages. 1967. DM 12,– / $ 3.00

Vol. 43: D. G. Quillen, Homotopical Algebra VI, 157 pages. 1967. DM 14,– / $ 3.50

Vol. 44: K. Urbanik, Lectures on Prediction Theory IV, 50 pages. 1967. DM 5,80 / $ 1.45

Vol. 45: A. Wilansky, Topics in Functional Analysis VI, 102 pages. 1967. DM 9,60 / $ 2.40

Vol. 46: P. E. Conner, Seminar on Periodic Maps IV, 116 pages. 1967. DM 10,60 / $ 2.65

Vol. 47: Reports of the Midwest Category Seminar I. IV, 181 pages. 1967. DM 14,80 / $ 3.70

Vol. 48: G. de Rham, S. Maumary et M. A. Kervaire, Torsion et Type Simple d'Homotopie. IV, 101 pages. 1967. DM 9,60 / $ 2.40

Vol. 49: C. Faith, Lectures on Injective Modules and Quotient Rings. XVI, 140 pages. 1967. DM 12,80 / $ 3.20

Vol. 50: L. Zalcman, Analytic Capacity and Rational Approximation, VI, 155 pages. 1968. DM 13.20 / $ 3.40

Vol. 51: Séminaire de Probabilités II. IV, 199 pages. 1968. DM 14,– / $ 3.50

Vol. 52: D. J. Simms, Lie Groups and Quantum Mechanics. IV, 90 pages. 1968. DM 8,– / $ 2.00

Vol. 53: J. Cerf, Sur les difféomorphismes de la sphère de dimension trois ($\Gamma_4 = O$). XII, 133 pages. 1968. DM 12,– / $ 3.00

Vol. 54: G. Shimura, Automorphic Functions and Number Theory VI, 69 pages. 1968. DM 8,– / $ 2.00

Vol. 55: D. Gromoll, W. Klingenberg und W. Meyer Riemannsche Geometrie im Großen VI, 287 Seiten. 1968. DM 20,– / $ 5.00

Bitte wenden / Continued

Vol. 56: K. Floret und J. Wloka,
Einführung in die Theorie der lokalkonvexen Räume
VIII, 194 Seiten. 1968. DM 16,– / $ 4.00

Vol. 57: F. Hirzebruch und K. H. Mayer,
O(n)-Mannigfaltigkeiten, exotische Sphären und Singularitäten.
IV, 132 Seiten. 1968. DM 10,80 / $ 2.70

Vol. 58: Kuramochi Boundaries of Riemann Surfaces.
IV, 102 pages. 1968. DM 9,60 / $ 2.40

Vol. 59: K. Jänich, Differenzierbare G-Mannigfaltigkeiten.
VI, 89 Seiten. 1968. DM 8,– / $ 2.00

Vol. 60: Seminar on Differential Equations and Dynamical Systems. Edited by G. S. Jones
VI, 106 pages. 1968. DM 9,60 / $ 2.40

Vol. 61: Reports of the Midwest Category Seminar II.
IV, 91 pages. 1968. DM 9,60 / $ 2.40

Vol. 62: Harish-Chandra, Automorphic Forms on Semisimple Lie Groups
X, 138 pages. 1968. DM 14,– / $ 3.50

Vol. 63: F. Albrecht, Topics in Control Theory.
IV, 65 pages. 1968. DM 6,80 / $ 1.70

Vol. 64: H. Berens, Interpolationsmethoden zur Behandlung von Approximationsprozessen auf Banachräumen.
VI, 90 Seiten. 1968. DM 8,– / $ 2.00

Vol. 65: D. Kölzow, Differentiation von Maßen.
XII, 102 Seiten. 1968. DM 8,– / $ 2.00

Vol. 66: D. Ferus, Totale Absolutkrümmung in Differentialgeometrie und -topologie. VI, 85 Seiten. 1968. DM 8,– / $ 2.00

Vol. 67: F. Kamber and P. Tondeur, Flat Manifolds.
IV, 53 pages. 1968. DM 5,80 / $ 1.45

Vol. 68: N. Boboc et P. Mustatà, Espaces harmoniques associés aux opérateurs différentiels linéaires du second ordre de type elliptique.
VI, 95 pages. 1968. DM 8,60 / $ 2.15

Vol. 69: Seminar über Potentialtheorie.
Herausgegeben von H. Bauer.
VI, 180 Seiten. 1968. DM 14,80 / $ 3.70

Vol. 70: Proceedings of the Summer School in Logic.
Edited by M. H. Löb.
IV, 331 pages. 1968. DM 20,– / $ 5.00

Vol. 71: Séminaire Pierre Lelong (Analyse), Année 1967-1968.
VI, 19 pages. 1968. DM 14,– / $ 3.50

Vol. 72: The Syntax and Semantics of Infinitary Languages.
Edited by J. Barwise.
IV, 268 pages. 1968. DM 18,– / $ 4.50

Vol. 73: P. E. Conner, Lectures on the Action of a Finite Group.
IV, 123 pages. 1968. DM 10,– / $ 2.50

Vol. 74: A. Fröhlich, Formal Groups.
IV, 140 pages. 1968. DM 12,– / $ 3.00

Vol. 75: G. Lumer, Algèbres de fonctions et espaces de Hardy. VI, 80 pages. 1968. DM 8 – / $ 2.00

Vol. 76: R. G. Swan, Algebraic K-Theory.
IV, 262 pages. 1968. DM 18,– / $ 4.50

Vol. 77: P.-A. Meyer, Processus de Markov: la frontière de Martin. IV, 123 pages. 1968. DM 10,– / $ 2.50

Vol. 78: H. Herrlich, Topologische Reflexionen und Coreflexionen.
XVI, 166 Seiten. 1968. DM 12,- / $ 3.00

Vol. 79: A. Grothendieck, Catégories Cofibrées Additives et Complexe Cotangent Relatif.
IV, 167 pages. 1968. DM 12,– / $ 3.00

Vol. 80: Seminar on Triples and Categorical Homology Theory. Edited by B. Eckmann
IV, 398 pages. 1969. DM 20,– / $ 5.00

Vol. 81: J.-P. Eckmann et M. Guenin, Méthodes Algébriques en Mécanique Statistique.
VI, 131 pages. 1969. DM 12,– / $ 3.00

Vol. 82: J. Wloka, Grundräume und verallgemeinerte Funktionen
VIII, 131 Seiten. 1969. DM 12,– / $ 3.00

Vol. 83: O. Zariski, An Introduction to the Theory of Algebraic Surfaces.
IV, 100 pages. 1969. DM 8,– / $ 2.00

Vol. 84: H. Lüneburg, Transitive Erweiterungen endlicher Permutationsgruppen.
IV, 119 Seiten. 1969. DM 10,– / $ 2.50

Vol. 85: P. Cartier et D. Foata, Problèmes combinatoires de commutation et réarrangements.
IV, 88 pages. 1969. DM 8,–/$ 2.00

Vol. 86: Category Theory, Homology Theory and their Applications I. Edited by P. Hilton.
VI, 216 pages. 1969. DM 16,–/$ 4.00

Vol. 87: M. Tierney, Categorical Constructions in Stable Homotopy Theory.
IV, 65 pages. 1969. DM 6,–/$ 1.50

Vol. 88: Séminaire de Probabilités III.
IV. 229 pages. 1969. DM 18,–/$ 4.50

Vol. 89: Probability and Information Theory.
Edited by M. Behara, K. Krickeberg and J. Wolfowitz.
IV, 256 pages. 1969. DM 18,–/$ 4.50

Vol. 90: N. P. Bhatia and O. Hajek, Local Semi-Dynamical Systems.
II, 157 pages. 1969. DM 14,–/$ 3.50

Vol. 91: N. N. Janenko, Die Zwischenschrittmethode zur Lösung mehrdimensionaler Probleme der mathematischen Physik.
VIII, 194 Seiten. 1969. DM 16,80 / $ 4.20

Vol. 92: Category Theory, Homology Theory and their Applications II. Edited by P. Hilton
V, 308 pages. 1969. DM 20,– / $ 5.00

Vol. 93: K. R. Parthasarathy, Multipliers on Locally Compact Groups.
III, 54 pages. 1969. DM 5,60 / $ 1.40

Vol. 95: A. S. Troelstra, Principles of Intuitionism.
II, 111 pages. 1969. DM 10,– / $ 2.50